职业教育旅游与餐饮类专业系列教材

烹饪英语

主　编　林　璐　谭艳红
副主编　麦　芳　张　娟　阳元妹　黄英初
参　编　卜嘉树　邓雪献　陈勇健　陆曼莎　蒋水秀
　　　　鲁　煊　李晓颖　谭顺捷　莫　强　赖亭君

机械工业出版社

本书以提升烹饪英语能力为出发点，根据烹饪行业的五大类术语（烹饪原料、烹饪方法、烹饪工具、烹饪调料和菜单）进行内容规划，整合为 17 个单元。全书分别从厨房工具和设备、早餐类、正餐类、食材类及食谱类知识进行单元内容设计，并将常见的食材词汇以图文及图表的形式呈现，为读者提供了帮助记忆的总结性资料，内容丰富实用，深度和广度适中。

本书强调口语的实践性，在深化烹饪英语词汇的基础上突出英语应用能力的培养，是职业院校、应用型本科院校各相关专业的适用教材，也可供餐饮从业者及烹饪爱好者自学使用。

本书配有电子课件等教师用配套教学资源，凡使用本书的教师均可登录机械工业出版社教育服务网 www.cmpedu.com 下载。咨询可致电：010-88379375，服务 QQ：945379158。

图书在版编目（CIP）数据

烹饪英语/林璐，谭艳红主编．—北京：机械工业出版社，2022.6（2024.2重印）

职业教育旅游与餐饮类专业系列教材

ISBN 978-7-111-70824-7

Ⅰ．①烹…　Ⅱ．①林…②谭…　Ⅲ．①烹饪—英语—职业教育—教材　Ⅳ．①TS972.1

中国版本图书馆CIP数据核字（2022）第086270号

机械工业出版社（北京市百万庄大街22号　邮政编码100037）

策划编辑：董宇佳　　责任编辑：董宇佳　张美杰

责任校对：炊小云　　责任印制：单爱军

北京虎彩文化传播有限公司印刷

2024年2月第1版第3次印刷

184mm×260mm・12.75印张・277千字

标准书号：ISBN 978-7-111-70824-7

定价：45.00元

电话服务	网络服务
客服电话：010-88361066	机　工　官　网：www.cmpbook.com
010-88379833	机　工　官　博：weibo.com/cmp1952
010-68326294	金　书　网：www.golden-book.com
封底无防伪标均为盗版	机工教育服务网：www.cmpedu.com

前言 Preface

越来越多的餐饮从业者意识到，随着我国国际影响力的显著提升，我国餐饮文化正在走向世界。餐饮从业人才的培养不仅要考虑面向市场、面向技术，也需要面向国际、面向未来。党的二十大报告首次将文明传播力、影响力与国际传播能力放在一起阐述，明确提出增强中华文明传播力影响力，坚守中华文化立场，讲好中国故事、传播好中国声音，展现可信、可爱、可敬的中国形象，推动中华文化更好走向世界。

我们的学生在未来的职业道路上不仅仅是在厨师岗位上发光发热，他们还将走向世界，立足于传播中国本土饮食文化，在服务的过程中将中国文化、中华美食和中国故事介绍给来自全世界的客人，结合"一带一路"倡议的推进，担负起传播中华优秀文化的责任。

本书针对大专院校旅游与餐饮类专业的学生进行编写，注重实用性，可减轻学生学习综合英语的负担，是一本简明、实用、形式活泼的烹饪英语教材。

烹饪英语涉及大量食材、烹饪制作及中西餐菜名的翻译以及表达，还包括烹饪行业术语，如烹饪原料、烹饪方法、烹饪工具和烹饪调料等，这些烹饪术语在烹饪英文资料和餐饮口语交际中大量使用，是烹饪英语课程的核心内容，因其多而广也是课程教学难点。

本书在编写过程中进行了精巧布置，由浅入深，注重实用性。

本书主要内容如下：

（1）第1部分"Getting into the Catering Business 初入餐饮业"。通过对点餐、厨师职位及厨房设备等相关内容的学习，为深入学习本课程奠定基础。

（2）第2部分"Breakfast 早餐"。掌握常见中西式早餐的英文表达，逐步了解中西饮食文化的差异。

（3）第3部分"Dinner 正餐"。从理解英文菜单上的词汇开始，逐步熟悉烹饪行业的术语。

（4）第4部分"Ingredients 食材"。了解关于火候、刀工、肉类食材和味道的英文表达，提高对菜名的中英文互译能力。此部分是学习本课程的重点。

（5）第5部分"Recipe 食谱"。理解英文食谱，掌握厨房工作交流常用功能句。此部分着重培养及强化学生烹饪英语的实际运用能力。

本书主要特色如下：

1. 书中的词汇、实用表达及对话都有对应的中文译文，针对学生发音困难的特点为大量词汇做了音标注释，并增添了大量的食材图片、总结性的词汇表格及菜式案例等，真正使学生可以在顶岗实习时即可在真实工作场景中加以运用，为学生之后的培训和工作打下基础。

2. 在每个单元的末尾，本书将"职业提示"融入了课程思政，致力于培养"英语+厨师"

的复合型技术技能人才。

  本书从约稿、定稿至出版，得到了校领导、各校同行、中国烹饪大师鲁煊以及广西烹饪餐饮行业协会会员单位荔园山庄的大力支持与帮助，在此，对他们以及编辑团队的辛勤劳动和精益求精的工作态度表示衷心感谢。由于编者水平有限，若有错漏之处敬请读者多提宝贵意见，以便再版时补充更正。

  本书配有电子课件等教师用配套教学资源，凡使用本书的教师均可登录机械工业出版社教育服务网 www.cmpedu.com 下载。咨询可致电：010-88379375，服务 QQ：945379158。

编　者

二维码索引
QR Code Index

序号	名称	二维码	页码	序号	名称	二维码	页码
1	Unit 1 Ordering Food in a Restaurant 在餐厅点餐		2	10	Unit 10 Menu of Chinese Food 中餐菜单		106
2	Unit 2 Having a Conversation with the Chef 与厨师交谈		12	11	Unit 11 Translation Techniques of Chinese Food 中餐菜名翻译技巧		116
3	Unit 3 Starting to Know the Kitchen 走进厨房		20	12	Unit 12 Meats and Poultry 肉类		130
4	Unit 4 Kitchen Tools and Equipment 厨房工具和设备		30	13	Unit 13 Aquatic Products 水产品		141
5	Unit 5 Western Breakfast 西式早餐		44	14	Unit 14 Food Tastes and Textures 食物的味道及口感		153
6	Unit 6 Chinese Breakfast 中式早餐		54	15	Unit 15 Condiments 调味品		164
7	Unit 7 Fruit Juices and Desserts 果汁和甜点		63	16	Unit 16 Common Measure Words in the Recipe 食谱中常见的量词		177
8	Unit 8 Quick Western Meals 西餐简餐		78	17	Unit 17 Common Functional Expressions in the Kitchen 厨房常用的功能句		186
9	Unit 9 Menu of Western Food 西餐菜单		93				

目录 Contents

Preface 前言
QR Code Index 二维码索引

01 Part 1　Getting into the Catering Business　// 001
- Unit 1　Ordering Food in a Restaurant 在餐厅点餐　// 002
- Unit 2　Having a Conversation with the Chef 与厨师交谈　// 012
- Unit 3　Starting to Know the Kitchen 走进厨房　// 020
- Unit 4　Kitchen Tools and Equipment 厨房工具和设备　// 030

02 Part 2　Breakfast　// 043
- Unit 5　Western Breakfast 西式早餐　// 044
- Unit 6　Chinese Breakfast 中式早餐　// 054
- Unit 7　Fruit Juices and Desserts 果汁和甜点　// 063

03 Part 3　Dinner　// 077
- Unit 8　Quick Western Meals 西餐简餐　// 078
- Unit 9　Menu of Western Food 西餐菜单　// 093
- Unit 10　Menu of Chinese Food 中餐菜单　// 106

04 Part 4　Ingredients　// 115
- Unit 11　Translation Techniques of Chinese Food 中餐菜名翻译技巧　// 116
- Unit 12　Meats and Poultry 肉类　// 130
- Unit 13　Aquatic Products 水产品　// 141
- Unit 14　Food Tastes and Textures 食物的味道及口感　// 153

05 Part 5　Recipe　// 163
- Unit 15　Condiments 调味品　// 164
- Unit 16　Common Measure Words in the Recipe 食谱中常见的量词　// 177
- Unit 17　Common Functional Expressions in the Kitchen 厨房常用的功能句　// 186

Appendix 附录　Pronunciation and Phonetic Symbols 读音和音标　// 196

References 参考文献　// 197

Part 1
Getting into the Catering Business
初入餐饮业

Unit 1 Ordering Food in a Restaurant 在餐厅点餐 / 002

Unit 2 Having a Conversation with the Chef 与厨师交谈 / 012

Unit 3 Starting to Know the Kitchen 走进厨房 / 020

Unit 4 Kitchen Tools and Equipment 厨房工具和设备 / 030

Unit 1　Ordering Food in a Restaurant
在餐厅点餐

随着经济全球化发展，中国与世界各国的交流日益频繁。掌握餐饮的基础用语，不仅可以更好地从事餐饮服务，更重要的是可以自信地走出去，从而有机会去体验不同的饮食文化。

Learning Objectives 学习目标

❋ Memorize the basic terms of ordering a meal.
　记住点餐的基础词汇。
❋ Know the useful expressions and functional sentences of dinner.
　了解关于用餐的实用表达和句子。

扫码看视频

Basic Terms 基础词汇

冰水　　/ˌaɪst ˈwɔːtə(r)/　**iced water**

自来水　/ˈtæp wɔːtə(r)/　**tap water**

柠檬水　/ˌleməˈneɪd/　**lemonade**

可乐　　/kəʊk/　**coke**

咖啡　　/ˈkɒfi/　**coffee**

奶茶　　/mɪlk tiː/　**milk tea**

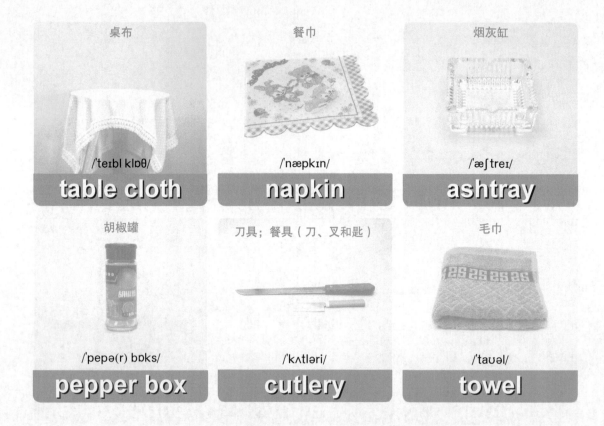

Functional Expressions 实用表达

请流利地朗读以下句子，并做到中英互译。

1. 欢迎光临我们餐厅！	Welcome to our restaurant!
2. 请问是两位顾客吗？	A table for two?
3. 请问有预订吗？	Do you have a reservation?
4. 请跟我来。	Follow me, please!
5. 坐在窗边可以吗？	Would you like to sit by the window?

welcome /ˈwelkəm/ v. 欢迎　　restaurant /ˈrestrɒnt/ n. 餐厅　　table /ˈteɪbl/ n. 桌子
reservation /ˌrezəˈveɪʃn/ n. 预订　　like /laɪk/ v. 喜欢　　sit /sɪt/ v. 坐下
window /ˈwɪndəʊ/ n. 窗户

6. 这个位置可以吗？	Is this table fine?
7. 请就座。	Be seated, please.
8. 这是菜单。	Here is the menu.
9. 先喝些什么？	What would you like to drink first?
10. 我想要一杯可乐。	I'd like a glass of coke.

seated /ˈsiːtɪd/ adj. 坐下的　　menu /ˈmenjuː/ n. 菜单
drink /drɪŋk/ v. 喝　　glass /ɡlɑːs/ n. 玻璃杯

11. 你想要点些什么？	What would you like, please?
12. 我要一份海鲜沙拉。	I would like to have the seafood salad.
13. 接下来点些什么？	What would you like to follow with?
14. 您的牛排要几成熟？	How would you like your steak?
15. 请稍等。	Just a moment, please.
16. 小心烫手。	Be careful. It's hot!

seafood /ˈsiːfuːd/ **n.** 海鲜　　salad /ˈsæləd/ **n.** 沙拉　　follow /ˈfɒləʊ/ **v.** 跟随
steak /steɪk/ **n.** 牛排　　moment /ˈməʊmənt/ **n.** 时刻，片刻　　careful /ˈkeəfl/ **adj.** 小心的

17. 我可以上菜了吗？	May I serve your dishes now?
18. 祝您用餐愉快。	Please enjoy your meal.
19. 祝您有个好胃口。	Have a good appetite.
20. 对不起，太太，我没听明白。	Pardon, madam. I'm afraid I didn't follow you.
21. 一切都还满意吗？	Is everything to your satisfaction?
22. 这是您的账单。	Here is the bill.

serve /sɜːv/ **v.** 为……服务　　dishes /ˈdɪʃɪz/ **n.** 菜（复数）　　enjoy /ɪnˈdʒɔɪ/ **v.** 享用
appetite /ˈæpɪtaɪt/ **n.** 胃口　　pardon /ˈpɑːdn/ **n.** 再说一次　　satisfaction /ˌsætɪsˈfækʃn/ **n.** 满意
bill /bɪl/ **n.** 账单

Task 任务　情景对话

Mr. White 来到餐厅用餐，请设计一段简单的对话。

时间：7 p.m. on Friday
地点：Diamond Restaurant
人物：客人 Mr. White，服务员 Johnny

Dialogues 对话

1. Seating Guests 领位

(服务员：Kevin，客人：Bella)

Kevin: Good day, sir and madam. Welcome to our restaurant!
Bella: Good day. A table for two, please.
Kevin: OK. Follow me, please. Mind your steps, please.
Bella: Thanks. We'd like to sit by the window, please.
Kevin: OK. Be seated, please. Here is the menu. Please take a look.
Bella: Thank you.

词汇

sir /sə(r)/ **n.** 先生
madam /ˈmædəm/ **n.** 女士
welcome /ˈwelkəm/ **v.** 欢迎
mind /maɪnd/ **v.** 留心
step /step/ **n.** 步伐，脚步
window /ˈwɪndəʊ/ **n.** 窗户
seated /ˈsiːtɪd/ **adj.** 坐下的
menu /ˈmenjuː/ **n.** 菜单

中文

Kevin：早上好，先生和女士。欢迎来到我们的餐厅！
Bella：早上好。我们有两个人。
Kevin：好的。请跟我来。请留心您的脚下。
Bella：谢谢。我们想坐在窗户旁边。
Kevin：好的。请坐。这是菜单，请看一看。
Bella：谢谢。

注释

（1）向他人问候。
Good morning! 早上好！
Good noon! 中午好！
Good afternoon! 下午好！
Good evening! 晚上好！
Good day! 日安！（白天见面或告别时的问候语）

（2）询问人数。
How many people, please? 请问有多少人？
A table for two? 两位吗？

（3）关于座位。
Where would you like to sit, please? 请问您想坐在哪儿？
Is this table fine? 这个位置好吗？
Have a seat, please. 请坐。

掌握英语音标是实现英语口语能力提升的关键。利用微信小程序功能，搜索"BBC 国际音标"，记住 48 个音标符号的发音。

2. Ordering Food 点菜

（服务员：Kevin，客人：Bella）

Kevin: Good day, sir.
Bella: Good day.
Kevin: What would you like, please?
Bella: I'd like a glass of coke, a hamburger and the fries.
Kevin: OK. Just a moment.

词汇

a glass of 一杯……
coke /kəuk/ *n.* 可乐
hamburger /ˈhæmbɜːgə(r)/ *n.* 汉堡包
fries /fraɪz/ *n.* 薯条
moment /ˈməʊmənt/ *n.* 片刻

中文

Kevin：你好，先生。
Bella：你好。
Kevin：请问您想要点什么？
Bella：我要一杯可乐、一个汉堡包还有薯条。
Kevin：好的，请稍等。

注释

（1）询问客人想点些什么。

Are you ready to order now, sir/madam? 准备好点餐了吗，先生/女士？
What would you like to have/order? 想点些什么？
Can I get you anything? Coffee? 想点些什么吗？咖啡？

（2）回答服务员。

I would like to have/order... 我想要……
A salad, please. 请给我一份沙拉。
I think I'll have a strawberry ice cream. 我想要一个草莓冰淇淋。
I'll just take a hot dog with extra cheese. 我要一个加奶酪的热狗。
Curried chicken for me, please. 我要一份咖喱鸡。

（3）英语有时通过词组来表示量词。

a slice of toast 一片吐司
a pot of tea 一壶茶
3 cans of coke 三罐可乐
a cup of coffee 一杯咖啡
2 glasses of red wine 两杯红酒

3. Paying Bills 结账

(服务员：Kevin，客人：Bella)

Kevin: How can I help you, madam?
Bella: I'd like to pay the bill.
Kevin: Just a moment, please. Here is the bill. Please check it.
Bella: Here is 100 yuan. Keep the change.
Kevin: Thanks. We look forward to serving you again.

词汇

bill /bɪl/ n. 账单
pay /peɪ/ v. 付款
check /tʃek/ v. 检查
100=one hundred /ˈhʌndrəd/ num. 一百
keep /kiːp/ v. 保有；留着；保持
change /tʃeɪndʒ/ n. 零钱
look forward to 期待
forward /ˈfɔːwəd/ adv. 向前
serve /sɜːv/ v. 为……服务
again /əˈgen/ adv. 再一次

中文

Kevin：夫人，我能为您做些什么吗？
Bella：我想买单。
Kevin：请等一下。这是账单，请检查一下。
Bella：这是 100 元。零钱不用找了。
Kevin：谢谢。我们期待再次为您服务。

注释

（1）询问是否需要服务。

How can I help you, sir/madam? 我能帮您什么忙吗，先生/女士？
May I help you? 请问需要帮忙吗？
What can I do for you? 我能帮您做些什么吗？

（2）买单的表达。

Bill, please. 请结账。
Check, please. 请结账。
I'd like to settle the bill, please. 我想结账。
Can you bring me the bill, please? 请把账单拿来好吗？

（3）与客人告别。

Welcome to come again. 欢迎下次光临。
Do come again, please. 欢迎再来。
Welcome to our restaurant next time. 欢迎再次光临我们餐厅。
Have a safe trip home. 一路平安。
Farewell! 再会！

练习：迎宾及点餐

厨师虽然不需要亲自服务客人，但是大家外出用餐时也时不时会遇见正与客人交流的厨师。后厨巡台，即指厨师主动跑到餐桌前去征询客人对菜品的意见，以便及时掌握客人反馈，改进菜品。这也是当下不少酒店正在开展的客户维护方式。厨师应该注重与客人之间的关系维护，可以学习一些简单的迎宾表达，让客人从进入餐厅时就能感受到来自厨师的诚挚欢迎。

1. 欢迎客人，并询问用餐人数

A. Good evening. Welcome to the restaurant. A table for two?
B. Welcome to come. How many persons, please?
C. Good day, sir and madam. Have you got a reservation?
D. We look forward to your next coming.

以上选项不正确的是：_____

2. 领位，并询问客人是否接受座位安排

A. Come with me, please. Would you prefer a table by the window?

B. Follow me, please. Where would you like to sit?

C. Come this way. Is this table fine?

D. Your table is not ready yet. Wait a moment in the lounge area, please.

以上选项不正确的是：_____

3. 将菜单递给客人

A. Here is the menu.

B. Here is the menu and drink list. Please take a look.

C. Here is our menu. Please have a look.

D. Here is the dessert.

以上选项不正确的是：_____

4. 为客人点饮料

A. What would you like to drink first?

B. While you look at the menu, may I bring you something to drink?

C. Anything to drink first? We have tea, coffee, and freshly squeezed fruit juices.

D. Would you like your coffee now or later?

以上选项不正确的是：_____

5. 给客人做决定的时间

A. I'll come back to take your order.

B. A waiter will come to take your orders.

C. The waiter/waitress will come back to take your orders. Just a moment, please.

D. You can enjoy the beautiful scenery outside the window.

以上选项不正确的是：_____

6. 询问客人是否可以点餐

A. Are you ready to order now?

B. May I take your order?

C. Would you like to order now?

D. We offer both Chinese and Western food.

以上选项不正确的是：_____

7. 询问客人需要什么头盘

A. What would you like for your appetizer?

B. What would you like for your starter?

C. Would you like a starter?

D. I'd like a salad.

以上选项不正确的是：_____

8. 询问客人要什么主菜

A. And for the main course?

B. What would you like for your main course?

C. What would you like for your entree?

D. Would you like something to eat?

以上选项不正确的是：_____

9. 询问客人是否需要甜点

A. Would you like some dessert?

B. Would you like to look at the dessert menu?

C. Anything for dessert?

D. What else would you like?

以上选项不正确的是：_____

10. 询问客人是否需要佐餐酒

A. Would you like to order some wine with your meal?

B. What wine would you like to go with your starter?

C. What would you like to drink with your steak?

D. Would you like to taste it?

以上选项不正确的是：_____

11. 询问客人关于某道菜式想要搭配的酱汁

A. What sauce would you like to go with your steak?

B. What kind of dressing would you like on your salad?

C. What sauce would you like spaghetti with?

D. Would you like to try a delicious dip?

以上选项不正确的是：_____

12. 询问客人还有什么需求

A. Is there anything else I can do for you?

B. Can I bring you anything else?

C. Do you have any special requirements?

D. What can I do for you?

以上选项不正确的是：_____

13. 询问客人对于这顿饭菜的体验

A. Was everything alright for you?

B. Is everything to your satisfaction?

C. How is everything going?

D. Can I assist you with something?

以上选项不正确的是：_____

14. 向客人表达欢迎下次光临

A. We're looking forward to having you again as our guest.

B. We hope to serve you again.

C. We look forward to serving you again.

D. You are welcome to come again.

以上选项正确的是（多选）：_____

职业提示

谦虚好学，善于沟通

中国式现代化是人口规模巨大的现代化，既需要农业、工业提供坚实的物质基础，也需要以更加优质高效的服务业托举起人民群众稳稳的幸福。厨师工作与餐厅服务息息相关，我们不仅要做好后厨的本职工作，还应了解餐厅服务的内容与流程；要善于沟通，很好地与团队成员合作共事，学会包容、尊重他人，以真诚热情的态度传递信息、交流情感，达到为客人提供高效服务的目的。

Unit 2 Having a Conversation with the Chef
与厨师交谈

在英语里，厨师有两个对应词汇：一个是"cook"，另一个是"chef"。在酒店和餐馆里，称呼主厨或专业厨师用"chef"，而称呼普通的厨师则使用"cook"。在客人用餐的过程中，"chef"偶尔有机会与客人面对面交流，讨论食材、口味、菜肴做法，抑或是处理关于食物的投诉，既维护了宾客关系，也能树立个人的专业与诚信形象。

Learning Objectives 学习目标

❋ Get familiar with expressions related to food comments.
  熟悉与食物品评相关的表达。
❋ Understand the concept of food preference.
  了解饮食偏好的概念。

扫码看视频

Basic Terms 基础词汇

菜肴 cuisine /kwɪˈziːn/

中餐 Chinese cuisine /ˌtʃaɪˈniːz kwɪˈziːn/

西餐 Western cuisine /ˈwestən kwɪˈziːn/

法国菜 French cuisine /frentʃ kwɪˈziːn/

特价菜 specials /ˈspeʃəlz/

每周特价 weekly specials /ˈwiːkli ˈspeʃəlz/

招牌菜 house specialty /haʊs ˈspeʃəlti/

本店特选酒 house wine /haʊs waɪn/

自制的 home-made /həʊm ˈmeɪd/

意大利面 pasta /ˈpæstə/

香味 aroma /əˈrəʊmə/

颜色 color /ˈkʌlə(r)/

形状 shape /ʃeɪp/

口感 taste /teɪst/

滋味 flavor /ˈfleɪvə(r)/

摆盘 presentation /ˌprizenˈteɪʃn/

食物偏好 food preference /fuːd ˈprefrəns/

主营 specialize in /ˈspeʃəlaɪz ɪn/

Functional Expressions 实用表达

请流利地朗读以下句子，并做到中英互译。

1. 这是我吃过最好吃的麻辣牛肉！ This is the best hot and spicy beef I've ever tasted!
2. 这道菜真让人惊艳！ This is really stunning!
3. 这家餐厅的川菜很美味。 This restaurant serves tasty Sichuan food.
4. 从食物上来看，食材都很新鲜。 Foodwise, the ingredients are all very fresh.
5. 我想见一见做这道菜的厨师。 I'd like to meet the chef who cooks this dish, please.

hot and spicy 麻辣，热辣
stunning /ˈstʌnɪŋ/ adj. 极好的
foodwise 从食物上看
spicy /ˈspaɪsi/ adj. 辣的
restaurant /ˈrestrɒnt/ n. 餐厅
ingredient /ɪnˈɡriːdiənt/ n. 原料
beef /biːf/ n. 牛肉
tasty /ˈteɪsti/ adj. 美味的
fresh /freʃ/ adj. 新鲜的

6. 你很会做菜。 You cook well.
7. 你做的牛排很好吃。 The steak you cooked is very delicious.
8. 香味让人垂涎三尺。 The smell makes my mouth watering.
9. 厨师，请在比萨上多撒一些罗勒。 Chef, please sprinkle more basil on top of the pizza.
10. 厨师，我的汤里面不要放葱。 Chef, I'd like my soup without any scallion, please.

steak /steɪk/ n. 牛排
mouth /maʊθ/ n. 嘴巴
basil /ˈbæzl/ n. 罗勒
soup /suːp/ n. 汤
delicious /dɪˈlɪʃəs/ adj. 美味的
watering /ˈwɔːtərɪŋ/ adj. 流口水
on top of... 在……上面
without /wɪˈðaʊt/ pre. 没有，不包含
smell /smel/ n. 味道
sprinkle /ˈsprɪŋkl/ v. 洒，撒
pizza /ˈpiːtsə/ n. 比萨
scallion /ˈskæliən/ n. 葱

11. 你有推荐吗？ What do you recommend?
12. 我想加热这道牛肉清汤。 I'd like to warm up this beef consommé.
13. 我们餐厅主营粤菜。 We specialize in Guangdong cuisine.
14. 这是本店招牌菜，试试看！ This is our house specialty. Give it a try!
15. 需要帮您端过去吗？ May I help you with this?

recommend /ˌrekəˈmend/ v. 推荐
beef consomme /biːf kənˈsɒmeɪ/ 牛肉清汤
cuisine /kwɪˈziːn/ n. 烹调法，菜肴
warm up 加热
specialize in /ˈspeʃəlaɪz ɪn/ 专门研究，专门经营
house specialty /haʊs ˈspeʃəlti/ 招牌菜

16. 请拿好。 Hold on to it, please.
17. 味道怎么样？ How does it taste?
18. 你最喜欢什么口味？ What is your favorite flavor?
19. 你吃饱了吗？ Have you had enough?
20. 感谢你的赞美。 Thanks for your compliments.

taste /teɪst/ v. 吃，尝
flavor /ˈfleɪvə/ n. 味道，风味
compliment /ˈkɒmplɪmənt/ n. 恭维，敬意，称赞
favorite /ˈfeɪvərɪt/ adj. 最喜欢的
enough /ɪˈnʌf/ adj. 足够的

Task 任务　情景对话

请设计一段简单的对话，向厨师表达你对食物的喜爱。

地点：Diamond Restaurant

人物：客人 Mr.White，行政副厨师长 Jackson

 Dialogues 对话

1. Compliments on Food 对食物的赞美

   (服务员：Kevin，客人：Bella)

Kevin: I'd recommend our Wellington steak, served with truffle and red wine sauce. Very delicious.
Bella: All right. I'll have that.
　　(...After finishing the main course, Bella gave a high praise for this dish.)
Bella: This is the best Wellington steak I've ever tasted! I can't praise this Wellington steak too highly.
Kevin: I'd be very glad you enjoy it! It's our sous chef Tom who made this. Just a moment, I'll let the chef know.
Bella: Great!

词汇

recommend /ˌrekəˈmend/ v. 推荐
Wellington steak 惠灵顿牛排
truffle /ˈtrʌfl/ n. 松露
red wine sauce 红酒酱汁
high praise /haɪ preɪz/ 高度赞扬，好评
sous chef /suːʃef/ 副厨师长

中文

Kevin：我推荐我们的惠灵顿牛排，搭配松露和红酒酱汁。非常美味。
Bella：好的。我就点这个了。
　　（……用完主菜后，贝拉对这道菜给予了高度赞扬。）
Bella：这是我吃过的最好的惠灵顿牛排！这个惠灵顿牛排怎么赞美都不过分。
Kevin：很高兴您喜欢这道菜！这是我们的副厨师长汤姆做的。请稍等一下，我转告厨师长。
Bella：太好了！

注释

（1）"我推荐"的表达。
I recommend + 菜名
I suggest + 菜名
*recommend 的语气相比 suggest 更强
如：I recommend/suggest strawberry ice cream. 我推荐草莓冰淇淋。
（2）给予高度好评 give a high praise。
因……受到极大好评：主语 + receive high praise for + 要表扬的对象
如：We received high praise for our house specialty. 我们的招牌菜受到了极大的好评。

(3)"转告"的表达。
I will let her know. 我会转告她。
I will tell him. 我会转告他。
I'll give him the message. 我会把消息转告他。
I'll pass your message on to her. 我会把你的口信转达给她。
I'll pass your kindness on to them. 我会向他们转达你的好意。

2. Complain About the Food 投诉食物

(客人：Bella，行政副厨师长：Jackson)

Bella:	Excuse me, I feel the steak is too salty, and it also tastes very strange.
Jackson:	I am very sorry. I will have the steak changed right away.
Bella:	Moreover, what I want is a medium steak, but this one is over-cooked.
Jackson:	I do apologize for this. Wait a moment, madam. This bottle of red wine is on the house, and your meal is free of charge tonight.
Bella:	That's good. Thanks.
Jackson:	I assure you it won't happen again.

词汇

feel /fiːl/ v. 觉得，感觉
steak /steɪk/ n. 牛排
salty /ˈsɔːlti/ adj. 咸的，含盐的
tastes /teɪsts/ v. 吃，尝（taste 的第三人称单数形式）
strange /streɪndʒ/ adj. 奇怪的
changed /tʃeɪndʒd/ v. 改变（change 的过去分词）
right away /raɪt əˈweɪ/ 立刻
moreover /mɔːrˈəʊvə(r)/ adv. 此外，而且
medium /ˈmiːdiəm/ adj. 五成熟的
over-cooked /ˈəʊvə(r) kʊkt/ adj. 煮过头的，煮糊的
apologize /əˈpɒlədʒaɪz/ v. 致歉
red wine /ˌred ˈwaɪn/ 红酒
on the house /ɒn ðə haʊs/ 免费，由店家出钱
free of charge /friː əv tʃɑːdʒ/ 免收费的
charge /tʃɑːdʒ/ n. 收费
assure /əˈʃʊə(r)/ v. 保证，担保
won't /wʌnt/ abbr. will not 的缩略形式

中文

Bella: 打扰一下，我觉得牛排太咸了，而且尝起来味道很奇怪。
Jackson: 非常抱歉。我马上为您换一份牛排。
Bella: 还有，我要的是五成熟的牛排，但是这份煮过头了。
Jackson: 我为此深表歉意。请稍等，夫人。
这瓶红酒由本店免费提供，还有您今晚的饭菜都是免费的。
Bella: 很好。谢谢。
Jackson: 我向你保证这种情况不会再发生了。

> **注释**
>
> （1）当客人不满意食物口味时的常见表达。
>
> | | |
> |---|---|
> | It's not to my taste. | 不合我的口味。 |
> | It's tasteless. | 没有味道。 |
> | It has a peculiar smell | 有奇怪的味道。 |
> | It's too greasy. | 菜很油腻。 |
> | I feel the dish is too salty/light/sour/bitter/spicy. | 我觉得这道菜太咸/淡/酸/苦/辣了。 |
>
> （2）客人对肉的熟度不满意时的常见表达。
>
> | | |
> |---|---|
> | The meat is over-cooked. | 肉煮过头了。 |
> | The meat is under-cooked. | 肉没有煮够时间。 |
> | The meat is raw. | 肉还是生的。 |
> | The meat is burnt. | 肉煮焦了。 |
>
> （3）客人对食材的新鲜度不满意时的常见表达。
>
> | | |
> |---|---|
> | It's not fresh. | 不新鲜。 |
> | The bread is stale. | 面包不新鲜了。 |
> | The chicken is moldy. | 鸡肉发霉了。 |
>
> 作为厨师，可提供的解决方法可以是重新做一道，或者额外提供菜式及给予免单的优惠。常见的表达是：
>
> We will have the dish changed right away. 我们马上为您更换。
> It's a complimentary dish from the chef. 这是厨师赠送的菜品。
> It's on the house. 由本店支付。
> It's free of charge. 不需支付。
> It is complimentary. 它是免费的。

3. Food Preference 对食物的偏好

(客人：Bella，行政副厨师长：Jackson)

词汇

home-made /həʊmˈmeɪd/ 自制的，家里做的
capellini /kæpəˈliːniː/ ***n.*** 天使细面（意大利面的一种）
lobster /ˈlɒbstə(r)/ ***n.*** 龙虾
sauce /sɔːs/ ***n.*** 酱汁
a bit /ə bɪt/ 有一点
longer /ˈlɒŋgə/ ***adj.*** 更久，更长
flavor /ˈfleɪvə(r)/ ***n.*** 风味，滋味
tender /ˈtendər/ ***adj.*** 嫩的，柔软的
moist /mɔɪst/ ***adj.*** 湿润的

Jackson: How is this home-made capellini with lobster sauce? I cook the lobster a bit longer than many chefs. It gives the dish more flavor and the meat is still tender and moist.

Bella: I like it. Pasta is my favorite type of cuisine! The combination of capellini and the sauce is wonderful. It's really delicious.
Jackson: Here is our weekly specials menu. Explore more pastas of our restaurant!
Bella: Fantastic. I'll go for pasta every time I order. Everyone has their preferences when it comes to food, you know.
Jackson: I got you. Come next time, please!

pasta /ˈpɑːstə/ **n.** 意大利面食	
favorite /ˈfeɪvərɪt/ **adj.** 最喜欢的	
type /taɪp/ **n.** 种类	
cuisine /kwɪˈziːn/ **n.** 菜肴；烹饪	
combination /ˌkɒmbɪˈneɪʃn/ **n.** 组合	
wonderful /ˈwʌndəf(ə)l/ **adj.** 极好的，绝妙的	
weekly specials /ˈwiːkli ˈspeʃəlz/ 每周特价	
restaurant /ˈrestrɒnt/ **n.** 餐厅	
fantastic /fænˈtæstɪk/ **adj.** 极好的；奇异的	
order /ˈɔːdə(r)/ **v.** 点餐（食物，饮料）	
preference /ˈprefrəns/ **n.** 偏爱；倾向	

中文

Jackson：这道自制的龙虾酱天使细面怎么样？和很多厨师比，我煮龙虾的时间会长些。这样更具风味，并且肉质仍然鲜嫩。
Bella：我喜欢。意大利面是我最喜欢的美食！天使细面和酱汁的搭配非常棒，非常好吃。
Jackson：这是我们的每周特价菜单。来发掘我们餐厅更多的意大利面吧！
Bella：太棒了。我每次点菜都会点意大利面。每个人在食物方面都有自己的喜好，你晓得。
Jackson：我懂。请一定下次再来！

注释

（1）饮食偏好。

1）sweet tooth 偏好甜食；prefer salty food 偏好咸的食品

例如：我喜欢甜食。

I have a sweet tooth.

2）我每次都点某一道菜。I'll go for _____ every time I order.

例如：我每次都点经典奶酪比萨。

I'll go for a classic cheese pizza every time I order.

3）他午餐只吃某一种食物。He sticks to _____ at lunch.

例如：我午餐只吃一份芝士三明治和一杯咖啡。

I stick to a cheese sandwich with a cup of coffee at lunchtime.

（2）home-made 在这里翻译为"本店自制的"。

home-made tofu 本店自制豆腐

home-made ice cream 本店自制冰淇淋

home-made wine 本店自制葡萄酒

> （3）combination 在这里翻译为"搭配"。
> The combination of strawberry syrup and orange juice makes the taste so good.
> 草莓糖浆和橙汁搭配的味道太好了。

练习：与客人进行简单交谈

根据提示的词汇及句型，尝试在以下情景中组织对话以进行应对。

1. Guests are not satisfied with the food. 客人对食物不满意。
A. What I want is a medium-rare steak, but this one is over-cooked. It also has a peculiar smell.
B. 抱歉：feel sorry; apologize; I'm so sorry that you have had such an unenjoyable evening.
    错误：mistake; error; fault
    再做一道：make another one; bring you a newly-made one
    重新做一道：re-cook this dish
    不合您的口味：not to your taste
    根据您的口味重新调味：season to your taste

2. Guests are unhappy about the speed of service. 客人对上菜的速度不满意。
A. The service is too slow! We waited so long for each course.
B. 抱歉让您久等了：sorry for the long wait; sorry to have kept you waiting
    人手不足：short of hands; short-handed; understaffed
    客人多：guests are crowded
    这是小小敬意：...with our compliments
    厨师赠送的甜点：complimentary dessert from the chef
    免费的：complimentary; on the house

3. Guests are very pleased with the taste of the food. 客人对食物的味道非常满意。
A. Great food and atmosphere here, we are very satisfied with our meal.
B. 很高兴您这么说：I'm glad you said so.
    我们选用了新鲜优质的食材：We use fresh and superior ingredients.
    给出你的建议：give us your opinion
    迎合我们客人的口味：cater to our guests' tastes
    我们以能够满足客人而自豪：We pride ourselves on satisfying our customers.
    期待你们的下一次光临：look forward to serving you again

4. Guests are curious about the tricks when cooking. 客人想知道食物的制作方法。
A. This is the best beef burger I have ever had! May I know what sauce did you use?
B. 这是我自己设计的配方：It's my own recipe.
    自制蛋黄酱：home-made mayonnaise

用香料提前腌制牛肉馅：use fresh spices to marinate the ground beef in advance
确保使用新鲜的原材料：make sure to use fresh food materials
用双份的芝士：use double cheese
选用柔软的面包：choose the soft and fluffy bun

职业提示

感恩社会，爱岗敬业

坚持"以人民为中心""人民至上"的原则。作为厨师，当客人提出意见时，我们首先要倾听客人的需求，换位思考，充分考虑客人的感受和想法，尊重顾客，不与其争吵。解决问题时，态度真诚、平和冷静、反应迅速，按照解决投诉的流程开展对客服务。培养自己灵活把握服务场景、娴熟处理特殊服务问题的能力。我们既要总结解决客人投诉的方法，更要学习如何降低客户的投诉率。

Unit 3 Starting to Know the Kitchen
走进厨房

餐厅端出的一道道的菜品就是厨房里所有工作者的劳动成果。比如，冷房出品的沙拉、热厨制作的牛排，以及饼房出品的甜点。在英语里，不同的部门及岗位都有不同的称呼。

Learning Objectives 学习目标

❋ Memorize the basic terms of kitchen titles.
记住厨房职位的基本术语。

❋ Grasp the useful expressions and functional sentences about kitchens.
掌握关于厨房的实用的表达和句子。

扫码看视频

Basic Terms 基础词汇

行政总厨 executive chef /ɪɡˈzekjətɪv ʃef/

行政副厨 executive sous chef /ɪɡˈzekjətɪv suː ʃef/

厨师长 chef /ʃef/

副厨师长 sous chef /suː ʃef/

主管 chef de partie /ʃef di pɑːti/

领班 demi chef de partie /demi ʃef di pɑːti/

帮厨 commis /ˈkɒmi/

学徒 apprentice /əˈprentɪs/

厨师 cook /kʊk/

热厨 hot kitchen /hɒtˈkɪtʃɪn/

冷厨 cold kitchen /kəʊld ˈkɪtʃɪn/

饼房 pastry kitchen /ˈpeɪstri ˈkɪtʃɪn/

冷柜 cooler /ˈkuːlə(r)/

冷藏室 freezer /ˈfriːzər/

肉类处理 butchery /ˈkʊtʃəri/

仓库 storage /ˈstɔːrɪdʒ/

收货区 pick-up section /pɪk ʌp ˈsekʃn/

厨师长办公室 chef's office /ʃefs ˈɒfɪs/

煎蛋档 egg station /eɡ steɪʃ(ə)n/

煮面档 noodle station /nuːdl steɪʃ(ə)n/

煎扒档 grill station /ɡrɪl steɪʃ(ə)n/

Functional Expressions 实用表达

请流利地朗读以下句子,并做到中英互译。

1. 行政总厨负责管理整个厨房。	The executive chef is in charge of the entire kitchen.
2. 行政副厨将负责指导培训工作。	The executive sous chef will be responsible for directing training.
3. 副厨师长会一直在厨房里。	Sous chef is always in the kitchen.
4. 菜单是厨师长制定的。	The chef prepares menus.
5. 厨师长会在装碟后及服务前试菜。	The chef will taste cooked foods before plate-up and service.

executive chef /ɪɡˈzekjətɪv ʃef/ 行政总厨
responsible /rɪˈspɒnsəbl/ **adj.** 负责的
plate /pleɪt/ **v.** 摆盘
sous chef /ˈsuː ʃef/ 副厨师长
taste /teɪst/ **v.** 吃,尝
service /ˈsɜːvɪs/ **n.** 服务

6. 主管负责厨房某个档口的工作。	A chef de partie is in charge of a particular station in the kitchen.
7. 在明档提供自助早餐。	The open kitchen offers breakfast buffets.
8. 我们饼房厨师长制作舒芙蕾非常拿手。	Our pastry chef is very good at soufflés.
9. 最有经验的厨师往往在热厨工作。	The most experienced cooks tend to work at the sauté station.
10. 一线厨师是真正为你做饭的人。	The line cooks are the people who actually cook your food.

chef de partie /ʃef di pɑːtiː/ 主管
particular /pəˈtɪkjələr/ **adj.** 特定的
offer /ˈɒfər/ **v.** 提供
buffet /ˈbʌfeɪ/ **n.** 自助餐
soufflé /ˈsuːfleɪ/ **n.** 舒芙蕾,蛋奶酥
sauté /ˈsəʊteɪ/ **v.** 炒,煎
in charge of 负责
station /ˈsteɪʃn/ **n.** 站点
breakfast /ˈbrekfəst/ **n.** 早餐
pastry chef /ˈpeɪstri ʃef/ 饼房厨师长
experienced /ɪkˈspɪəriənst/ **adj.** 熟练的

11. 我们的第一份工作通常都是初级厨师。	Normally our first jobs are commis.
12. 即使是煎蛋也需要具有基本的烹饪知识。	To fry an egg also requires basic cooking knowledge.
13. 我们先来煮汤吧。	Let's get going on the soup.
14. 要把食品安全放在首位。	Food safety should be a top priority.
15. 厨房每天都要仔细清洁。	The kitchen must be cleaned carefully every day.

commis /ˈkɒmi/ **n.** 初级厨师，帮厨
basic /ˈbeɪsɪk/ **adj.** 基本的
soup /suːp/ **n.** 汤
top /tɒp/ **adj.** 最高的
carefully /ˈkeəfəli/ **adv.** 仔细地

fry /fraɪ/ **v.** 煎，炒
cooking /ˈkʊkɪŋ/ **adj.** 烹调的
food /fuːd/ **n.** 食物
priority /praɪˈɒrəti/ **n.** 优先权

require /rɪˈkwaɪə(r)/ **v.** 要求，需要
knowledge /ˈnɒlɪdʒ/ **n.** 知识
safety /ˈseɪfti/ **n.** 安全
clean /kliːn/ **v.** 打扫

16. 检查厨房的每一个角落以确保蟑螂都被清除。 | Check every corner to ensure cockroaches are removed from the kitchen.
17. 要更注意个人卫生。 | Be more aware of personal hygiene.
18. 餐具在使用以前要消毒。 | The tableware needs to be sterilized before use.
19. 把你的制服和围裙送到布草间即可。 | Just bring your uniform and apron to the linen room.

check /tʃek/ **v.** 检查
ensure /ɪnˈʃʊr/ **v.** 确保
remove /rɪˈmuːv/ **v.** 去除
tableware /ˈteɪblweər/ **n.** 餐具
uniform /ˈjuːnɪfɔːm/ **n.** 制服

corner /ˈkɔːnər/ **n.** 角落
cockroaches /ˈkɒkrəʊtʃɪz/ **n.** 蟑螂（复数）
personal hygiene /pɜːsənl haɪdʒiːn/ 个人卫生
sterilized /ˈsterəlaɪzd/ **adj.** 消过毒的
apron /ˈeɪprən/ **n.** 围裙

Task 任务　自我介绍

初入厨房时，学着向同事介绍自己。

地点：Kitchen

人物：初级厨师 Steve，厨师长 Jackson

Dialogues 对话

1. Getting into the Kitchen 走进厨房

（初级厨师：Steve，行政副厨师长：Jackson）

Jackson: Nice to meet you, Steve. My name is Jackson. I'm executive sous chef, responsible for your training. Let me give you a brief.

Steve: Thank you, chef. It's my honor.

Jackson: Here is the hot kitchen. It's also called sauté station. It's the place where the most experienced cooks tend to work at.

Steve: It's also my intent. I will make effort on this.

词汇

executive /ɪgˈzekjətɪv/ **adj.** 行政的
sous chef /ˈsuː ʃef/ 副厨师长
responsible /rɪˈspɒnsəbl/ **adj.** 负责的
honor /ˈɒnə(r)/ **n.** 荣誉，有幸
sauté station 西餐厨房里的炒菜档
experienced /ɪkˈspɪəriənst/ **adj.** 熟练的
cook /kʊk/ **n.** 厨师，厨子
tend to 倾向于
intent /ɪnˈtent/ **n.** 意图，目的
effort /ˈefət/ **n.** 努力

Jackson: Brilliant. But now you are assigned to work in the pastry and bakery kitchen. You will work under the pastry demi chef, to assist ingredient preparations.

Steve: I understand, chef. Thank you for giving me this opportunity.

brilliant /ˈbrɪliənt/ *adj.* 极棒的
pastry /ˈpeɪstri/ *n.* 油酥，点心
bakery /ˈbeɪkəri/ *n.* 面包店，烘焙房
demi chef /demi ʃef/ 领班（demi chef 是 demi chef de partie 的缩略表达）
assist /əˈsɪst/ *v.* 协助
ingredient /ɪnˈɡriːdiənt/ *n.* 原料
preparation /ˌprepəˈreɪʃn/ *n.* 准备
understand /ˌʌndəˈstænd/ *v.* 理解，懂
opportunity /ˌɒpəˈtjuːnəti/ *n.* 机会

中文

Jackson：很高兴认识你，史蒂夫。我叫杰克逊。我是行政副厨，负责你的培训。让我简要介绍一下。

Steve：谢谢厨师长。这是我的荣幸。

Jackson：这是热厨，也称为炒站。这往往是经验最丰富的厨师倾向于工作的地方。

Steve：这也是我的打算。我会为此努力的。

Jackson：很棒。但是现在您被分配到饼房工作。你将在饼房领班手下干活，协助准备配料。

Steve：我明白，厨师长。感谢您给我这次机会。

注释

（1）厨房职位的词汇。

executive chef /ɪɡˈzekjətɪv ʃef/	行政总厨
executive sous chef /ɪɡˈzekjətɪv suː ʃef/	行政副厨
chef /ʃef/	厨师长
sous chef /ˈsuː ʃef/	副厨师长
chef de partie /ʃef di pɑːti/	主管
demi chef de partie /ˈdemi ʃef di pɑːti/	领班
commis /ˈkɒmi/	初级厨师，帮厨
apprentice /əˈprentɪs/	学徒

（2）厨房结构常用词汇。

hot kitchen /hɒt ˈkɪtʃɪn/	热厨
cold kitchen /kəʊld ˈkɪtʃɪn/	冷厨
pastry kitchen /ˈpeɪstri ˈkɪtʃɪn/	饼房

注释

fish section /fɪʃ 'sekʃn/	水台
butchery /'bʊtʃəri/	肉类处理
storage /'stɔːrɪdʒ/	仓库
freezer /'friːzər/	冷藏室
beverage cooler /'bevərɪdʒ kuːlər/	饮料冷柜
scullery /'skʌləri/	碗碟洗涤室
chef's office /ʃefs 'ɒfɪs/	厨师长办公室
pick-up section /pɪk ʌp 'sekʃn/	收货区

（3）中餐厨房的组织结构。

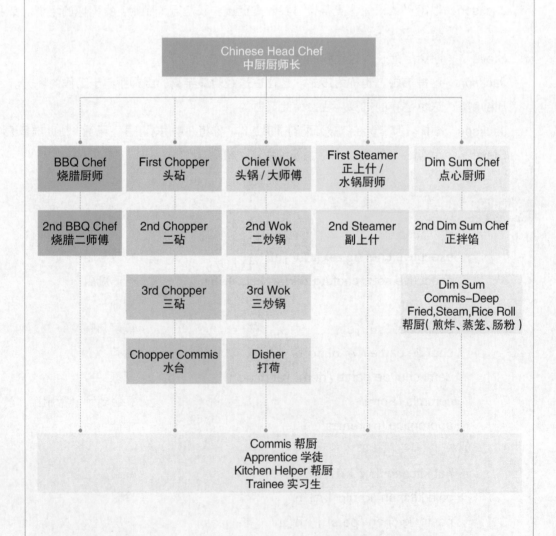

（4）西餐厨房的组织结构。

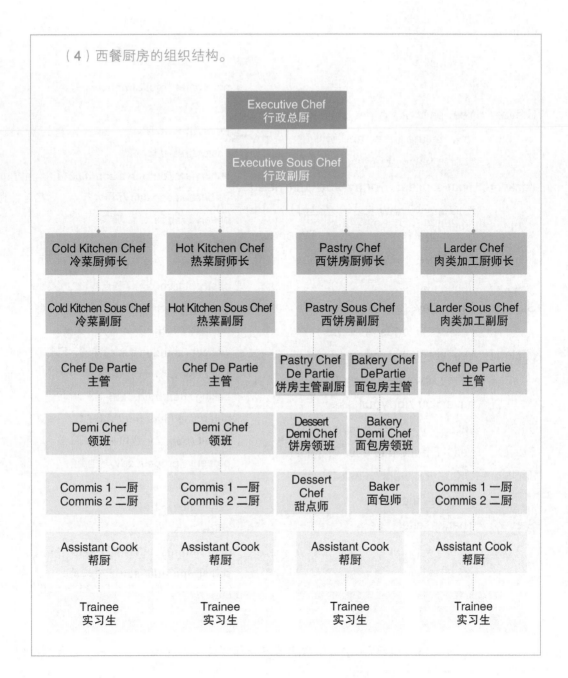

2. Learning to Ask Questions 学习向他人提问

（初级厨师：Steve，行政副厨师长：Jackson）

Jackson: As a commis chef, you will be working under pastry demi chef—Rosa. She is an excellent baker, also a good teacher.

词汇

commis /ˈkɒmi/ *n.* 帮厨
under /ˈʌndə(r)/ *prep.* 在……之下
pastry demi chef 饼房领班
excellent /ˈeksələnt/ *adj.* 优秀的
baker /ˈbeɪkə(r)/ *n.* 面包师

Steve:	Yes, chef. As a newcomer, I will try to fit in the team and will assist my supervisor as best as I can.
Jackson:	That's great. Today you just need to observe how to set up and break down the pastry station. Whenever you have queries, don't hesitate to ask.
Steve:	Um, I have one question. I wonder if I may test foods after baking?
Jackson:	Yes, you may. You need to determine if they have been baked properly. Oh, have you got your health certificate yet?
Steve:	Yes. I'll bring it to your office later.
Jackson:	OK. If you want to ask things politely, just use "may I ask...".
Steve:	Thanks, chef. I see.

词汇

newcomer /ˈnjuːkʌmə(r)/ **n.** 新人
fit in /fɪt ɪn/ 适应
team /tiːm/ **n.** 团队
assist /əˈsɪst/ **v.** 协助
supervisor /ˈsuːpərvaɪzər/ **n.** 监督人，主管
as best as one can 尽最大努力；竭尽全力地
observe /əbˈzɜːv/ **v.** 观察
set up 安装，装置好
break down 分解，破除
pastry station 点心 / 糕点档口
query /ˈkwɪəri/ **n.** 疑问
hesitate /ˈhezɪteɪt/ **v.** 犹豫，迟疑
wonder /ˈwʌndə(r)/ **v.** 想知道，好奇
test /test/ **v.** 测试，测验
baking /ˈbeɪkɪŋ/ **n.** 烘焙
determine /dɪˈtɜːmɪn/ **v.** 确定
baked /beɪkt/ **v.** 烘烤
properly /ˈprɒpəli/ **adv.** 恰当地
health /helθ/ **n.** 健康
certificate /səˈtɪfɪkət/ **n.** 证书
office /ˈɒfɪs/ **n.** 办公室
later /ˈleɪtə(r)/ **adv.** 稍后，晚些
politely /pəˈlaɪtli/ **adv.** 有礼貌地

中文

Jackson：	作为初级厨师，你将在饼房领班厨师罗莎手下工作。她是一位出色的面包师，还是一位好老师。
Steve：	好的，厨师长。作为一个新人，我会努力融入团队并尽我所能为我的主管提供帮助。
Jackson：	太好了。今天你只需要观察如何摆好及收拾甜点档。如有任何疑问，随时提出。
Steve：	呢，我有一个问题。我想知道我是否可以在食物烘焙好后测试品尝？
Jackson：	可以。你需要确定它们是否烘焙得当。噢，你获得健康证了吗？
Steve：	是的。我稍后再带到您的办公室。
Jackson：	好的。如果您想礼貌地问问题，只需使用"我可以问……吗"。
Steve：	谢谢，厨师长。我懂了。

> **注释**
>
> （1）supervisor 监督人，也可以译为：导师、主管、上级。
>
> （2）as best as I can 意为"我尽力做好（这件事）"。以下是几个"as...as..."句型或短语：
>
> > as soon as possible 尽快
> >
> > as fast as someone 与某人一样快
> >
> > as long as 只要……（就……）
> >
> > as well as someone do 做得和某人一样好
>
> （3）station 在这里意为"档口"，比如在餐厅里还有：
>
> > noodle station 面档
> >
> > egg station 煎蛋档
> >
> > congee station 粥档
> >
> > BBQ(barbeque) station 烧烤档

3. Being Guided 得到指引

（初级厨师：Steve，饼房领班：Rosa）

Rosa: Steve, wash and peel some fresh fruits and vegetables, please.

Steve: Yes, chef. Right away. I have just finished disinfecting the tables and knives.

Rosa: Let me check it. Yes, you've done a good job. Keeping a clean workspace is essential for this job.

Steve: Thank you, chef. And what to do after the fruits and vegetables?

Rosa: No rush. Focus on one task at a time, minimize distractions. I will tell you then.

> **词汇**
>
> wash /wɒʃ/ **v.** 洗
> peel /piːl/ **v.** 去皮
> fresh fruit /freʃ fruːt/ 鲜果
> vegetable /ˈvedʒtəb(ə)l/ **n.** 蔬菜
> right away 即刻，马上
> disinfecting /ˌdɪsɪnˈfektɪŋ/ **v.** 消毒（disinfect 的 ing 形式）
> tables /ˈteɪblz/ **n.** 桌子（复数）
> knives /naɪvz/ **n.** 刀子，刀具（复数）
> keeping /ˈkiːpɪŋ/ **v.** 保持（keep 的 ing 形式）
> clean /kliːn/ **adj.** 干净的
> workplace /ˈwɜːkpleɪs/ **n.** 工作场所
> essential /ɪˈsenʃl/ **adj.** 必不可少的，绝对必要的
> after /ˈɑːftə(r)/ **prep.** 在……之后
> focus on /ˈfəʊkəs ɒn/ 集中于
> task /tɑːsk/ **n.** 任务
> minimize /ˈmɪnɪmaɪz/ **v.** 最小化
> distraction /dɪˈstrækʃn/ **n.** 注意力分散，分心，干扰
> then /ðen/ **adv.** 那时，介时

中文

Rosa：史蒂夫，把这些新鲜的水果和蔬菜洗净并去皮。

Steve：好的，厨师长。马上。我刚刚完成了餐桌和刀的消毒工作。

Rosa：我检查一下。好，你干得不错。保持工作场所干净对于这项工作很关键。

Steve：谢谢，厨师长。洗完水果和蔬菜之后该做什么？

Rosa：不要着急。一次专注于一项任务，尽量减少干扰。到时我再告诉你。

注释

（1）you've done a good job 你做得很好，你干得不错。

"做得好"的表达还有：well done, good work, a bang-up job

"做得好，继续保持"：keep up the good work

如：Steve did a bang-up job with the food presentation. 斯蒂夫的摆盘非常出色。

Keep up the good work. You are the best candidate for sous chef. 继续努力，你是副厨的最佳人选。

（2）no rush 不急，别着急。

"不要急"的表达还有：don't rush, no hurry, no hurries, take your time, take it easy

如：I'm in no hurry to get things done. 我不急着把事情做完。

You can finish your job before 7. pm, so there's no rush. 你可以在晚上7点前完成工作，所以不必着急。

（3）distraction 干扰，分心。

如：It was difficult to work with so many distractions. 有这么多让人分心的事，工作真不容易。

Do your job without any distractions. 不要分心，做好你的工作。

（4）"到时再说"的表达。

We will talk about it then. 我们到时候再说。

Let's talk about it later. 我们晚些时候再说吧。

练习：熟悉厨房常见职位名词

将左侧职位名词翻译为英文，写在右侧方框内并背诵。

职业提示

价值引领，人生情怀

通过对厨房部门和岗位的了解，同学们首先要意识到厨房工作责任重大，关系到顾客的安全和健康。同时厨房工作辛苦繁重，需要我们认真、细心，全力以赴。从古至今的中国烹饪大师们无不是凭着满腔热情，刻苦钻研烹饪技术，才在专业上有所建树的。同学们需要在学习之初就建立职业认同感，做好自己的职业生涯规划。青年强，则国家强。当代中国青年生逢其时，施展才干的舞台无比广阔，实现梦想的前景无比光明。

Unit 4　Kitchen Tools and Equipment
厨房工具和设备

厨房里的器具繁多，用途不一。掌握这些常用工具的英文词汇，既有助于与外籍同事的交流，也有助于读懂英文食谱及自学英文网站上的学习资料。想象一下，当有一天可以和不同国籍的厨师们一起工作，互相学习彼此的烹饪方法，要多谢自己现在已经将厨房工具的词汇提前学会了哦！

Learning Objectives 学习目标

❋ Memorize the words and expressions of utensils and equipment in the kitchen.
　记住厨房用具和设备的单词和表达。

❋ Learn new phrases and sentences of food processing.
　掌握食物加工的新短语和句子。

扫码看视频

Basic Terms 基础词汇

Unit 4　Kitchen Tools and Equipment　厨房工具和设备

隔水蒸锅	炖锅	炖菜锅

/ˌbæn məˈriː/	/ˈstjuːpæn/	/ˈbreɪzər pæn/
bain marie	**stewpan**	**braiser pan**

电饭锅	烤箱	烟熏炉
/raɪsˈkʊkə(r)/	/ˈʌvn/	/ˈsməʊkɪŋ ˈʌvn/
rice cooker	**oven**	**smoking oven**

电磁炉	微波炉	吸油烟机

/ɪnˈdʌkʃn ˈkʊkə(r)/	/ˈmaɪkrəweɪvˈʌvn/	/ˈventɪleɪtə(r)/
induction cooker	**microwave oven**	**ventilator**

燃气灶	烤炉	火炉

/gæs ˈbɜːnə(r)/	/grɪl/	/stəʊv/
gas burner	**grill**	**stove**

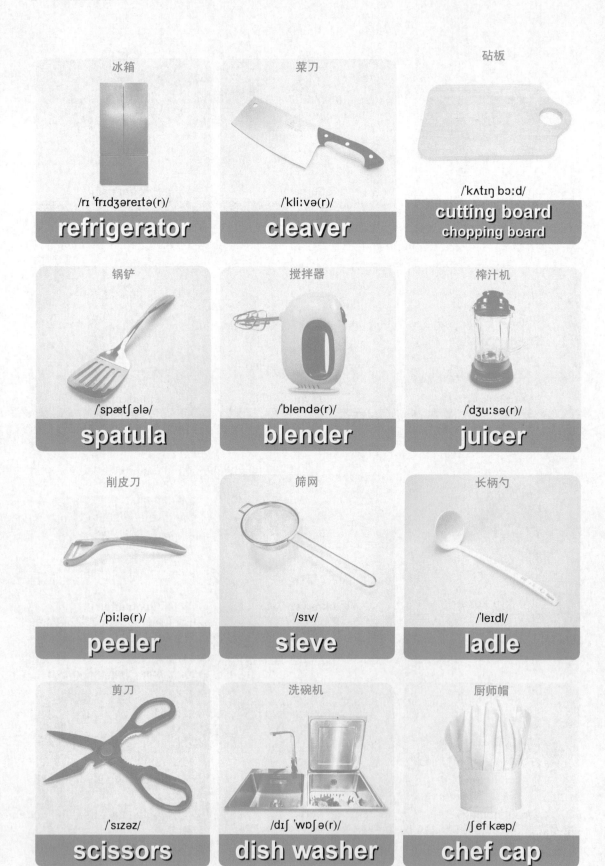

Functional Expressions 实用表达

请流利地朗读以下句子，并做到中英互译。

1. 我们厨房配备有齐全的炊具。 Our kitchen is well equipped with cookware.
2. 平底锅是西式厨房最常用的工具。 The pan is the most useful utensil in western kitchen.
3. 长柄煎锅是我们厨房不可或缺的工具。 Skillet is an indispensable tool in our kitchen.
4. 炒锅是做中餐的好工具。 Wok is a great tool when cooking Chinese cuisine.
5. 好的厨师从不埋怨工具。 A good cook never blames his tools.

equipped with /ɪˈkwɪpt wɪð/ 配备了
pan /pæn/ n. 平底锅
utensil /juːˈtensl/ n. 用具
indispensable /ˌɪndɪˈspensəbl/ adj. 不可缺少的
wok /wɒk/ n. 锅（源自粤语）
blame /bleɪm/ v. 责怪
cookware /ˈkʊkweə(r)/ n. 炊具
most useful /məʊst ˈjuːsfl/ 最有用的
skillet /ˈskɪlɪt/ n. 长柄平底煎锅
tool /tuːl/ n. 工具
Chinese cuisine /ˌtʃaɪˈniːz kwɪˈziːn/ 中餐

6. 去冰柜里把材料拿出来。 Bring the materials from the freezer.
7. 用沥水篮清洗并沥干蔬菜。 Use a strainer to rinse and drain the vegetable.
8. 用不粘锅做煎饼。 Use a non-stick pan to make the pancake.
9. 用抹刀将所有材料混合。 Mix all the ingredients with a spatula.
10. 把锅放在炉子上。 Put the pot on the stove.

material /məˈtɪəriəl/ n. 材料
strainer /ˈstreɪnə(r)/ n. 滤网，过滤器
drain /dreɪn/ v. 排水，沥水
non-stick /ˌnɒnˈstɪk/ adj. 不沾的
mix /mɪks/ v. 混合
spatula /ˈspætʃələ/ n. 抹刀，小铲
stove /stəʊv/ n. 火炉
freezer /ˈfriːzər/ n. 冰箱，冷冻库
rinse /rɪns/ v. 冲洗
vegetable /ˈvedʒtəb(ə)l/ n. 蔬菜
pancake /ˈpænkeɪk/ n. 煎饼
ingredient /ɪnˈɡriːdiənt/ n. 原料
pot /pɒt/ n. 锅，盆，罐

11. 点燃燃气灶。 Turn on the gas stove.
12. 把锅开大火，烧开。 Put the pan over high heat and bring it to a boil.
13. 把猪骨放入锅里，不加盖。 Put the pork bones into the pot without the lid on.
14. 把肉放进一个大炖锅里。 Put the meat into a large saucepan.
15. 盖好盖子。 Put the lid on.

turn on 打开，发动
bring it to a boil /brɪŋ ɪt tu ə bɔɪl/ 煮沸，烧开
meat /miːt/ n. 肉
gas stove 燃气灶
pork bone /pɔːk bəʊn/ 猪骨
saucepan /ˈsɔːspən/ n. 炖锅，深平底锅
high heat /haɪ hiːt/ 高温，大火
lid /lɪd/ n. 盖子

16. 平底锅盖上盖。	Cover the pan.
17. 把青椒放到烤架下。	Place the green peppers under the grill.
18. 把一半的配料撒在烤盘上。	Sprinkle half of the ingredients over the roasting pan.
19. 我们用烤箱做烧鸭。	The duck is roasted in the oven.
20. 时间到了。关掉烤箱。	Time is up. Turn off the oven.

cover /ˈkʌvər/ **v.** 覆盖，盖上
green pepper /ɡriːn ˈpepə(r)/ 青椒
sprinkle /ˈsprɪŋkl/ **v.** 撒，洒
roasting pan /ˈrəʊstɪŋ pæn/ 烤盘
roasted /ˈrəʊstɪd/ **v.** 热煨（roast 的过去分词）
turn off /tɜːn ɒf/ 关掉，关闭

place /pleɪs/ **v.** 放置
grill /ɡrɪl/ **n.** 烤炉，烤架
half of /hɑːf əv/ 一半的
duck /dʌk/ **n.** 鸭肉
oven /ˈʌvn/ **n.** 烤箱

Task 任务 口语训练

试着用一句含厨具的英文来表达自己要做的事情。
1）使用……清洗食材 rinsing and draining
2）使用……烹煮食物 cooking
地点：Kitchen
人物：初级厨师

 ### Dialogues 对话

1. Rinsing the Lettuce 洗生菜

（初级厨师：Steve，行政副厨师长：Jackson）

词汇

rinse /rɪns/ **v.** 清洗
lettuce /ˈletɪs/ **n.** 莴苣，生菜
knife /naɪf/ **n.** 刀
break /breɪk/ **v.**（使）破，裂，碎
root /ruːt/ **n.** 植物的根茎
off /ɒf/ **adv.** 脱离，脱掉，去除
cut off /ˈkʌt ɒf/ 切断，切除
wilted /ˈwɪltɪd/ **adj.** 枯萎的，萎蔫的
spot /spɒt/ **n.** 斑点，点，部位
combined /kəmˈbaɪnd/ **adj.** 结合的
baking soda /ˈbeɪkɪŋ səʊdə/ 小苏打

Jackson: Steve, I need you to rinse these lettuces. Use a knife to break the root off the lettuce. Then cut off any wilted spots.

Steve: Yes, chef. I will wash them in water combined with baking soda.

Jackson: That will be great! It helps to remove pesticide residue and surface dirt. Steve: How to deal with the lettuce after rinsing? Jackson: Just put the leaves into the colander. We want them for salad after draining. Steve: Yes, chef.	remove /rɪˈmuːv/ **v.** 去除，移走 pesticide /ˈpestɪsaɪd/ **n.** 杀虫剂 residue /ˈrezɪdjuː/ **n.** 残渣，剩余 surface /ˈsɜːfɪs/ **n.** 表面 dirt /dɜːt/ **n.** 污垢，泥土 deal with /diːl wɪð/ 处理 leaves /liːvz/ **n.** 叶子（leaf 的复数） colander /ˈkʌləndər/ **n.** 滤器，滤锅 salad /ˈsæləd/ **n.** 沙拉 draining /ˈdreɪnɪŋ/ **v.** 排水，沥干（drain 的现在分词）

中文

Jackson：史蒂夫，我需要你来处理这些生菜。用刀将生菜的根部割开。然后切掉所有枯萎的部位。

Steve：好的，厨师长。我打算用小苏打水来清洗。

Jackson：那太好了！它有助于去除农药残留和表面污垢。

Steve：洗好后如何处理这些生菜？

Jackson：把菜叶放到滤网里。把它们沥干后做沙拉。

Steve：好的，厨师长。

注释

（1）"wash"与"rinse"的区别：wash 意为洗、洗涤（通常使用洗涤剂），而 rinse 是指用清水冲洗、漂洗、洗涤。在表达用清水冲洗食材时，使用"rinse"一词。

（2）"lettuce"意为"莴苣，生菜"。常见的凯撒沙拉（caesar salad）里的生菜品种为长叶莴苣（romaine lettuce，来自法语，意为罗马生菜）；Chinese lettuce 意为莜麦菜。

（3）"break off" "cut off"里的"off"用来表示"除去了某物"，类似的表达还有：

    take off 脱下（衣帽等）

    bite off 咬掉，啃掉

    get off （从……）下来

    turn off 关掉（水源、火源、电源等）

2. The Knife, Chopping Board and Peeler 刀、砧板和削皮刀

(初级厨师：Steve，行政副厨师长：Jackson)

Jackson: Steve, you need to know how to deal with root vegetables. There are some common root vegetables, such as onions, carrots, turnips, potatoes, lotus roots, and pumpkins.

Steve: First up, I think we should clean them or rinse them properly. Next...just to cut them.

Jackson: Correct. Don't forget to sharpen your knife before every time you use it. Keep a sharp knife with the sharpening steel. It's dangerous working in a kitchen with a blunt knife.

Steve: I understand, chef. Then we start the chopping.

Jackson: Place a wet towel underneath the chopping board to stop it rocking or slipping. It's also necessary to use a vegetable peeler when slicing and shaving.

Steve: Yes, chef. I still have a lot to learn.

词汇

root vegetable /ruːt ˈvedʒtəbl/ 根茎类蔬菜
root /ruːt/ **n.** 根茎
common /ˈkɒmən/ **adj.** 常见的
onion /ˈʌnjən/ **n.** 洋葱
carrot /ˈkærət/ **n.** 胡萝卜
turnip /ˈtɜːnɪp/ **n.** 白萝卜
potato /pəˈteɪtəʊ/ **n.** 土豆
lotus root /ˈləʊtəs ruːt/ 莲藕
pumpkin /ˈpʌmpkɪn/ **n.** 南瓜
rinse /rɪns/ **v.** 冲洗
properly /ˈprɒpəli/ **adv.** 恰当地
cut /kʌt/ **v.** 切，割
correct /kəˈrekt/ **adj.** 正确的
sharpen /ˈʃɑːpən/ **v.** 使……变锋利
knife /naɪf/ **n.** 刀
sharp /ʃɑːp/ **adj.** 锋利的
sharpening steel /ˈʃɑːpnɪŋ stiːl/ 磨刀棒
dangerous /ˈdeɪndʒərəs/ **adj.** 危险的
blunt /blʌnt/ **adj.** 不锋利的，钝的
chopping /ˈtʃɒpɪŋ/ **v.** 砍，剁（chop 的现在分词）
place /pleɪs/ **v.** 放置
wet towel /wet ˈtaʊəl/ 湿毛巾
underneath /ˌʌndəˈniːθ/ **prep.** 在……底下
chopping board /ˈtʃɒpɪŋ bɔːrd/ 砧板
rocking /ˈrɒkɪŋ/ **v.** 摇摆，振动
slipping /ˈslɪpɪŋ/ **v.** 滑动
necessary /ˈnesəs(ə)ri/ **adj.** 必需的，必要的
vegetable peeler /ˈvedʒtəbl piːlə(r)/ 蔬菜削皮刀
slicing /ˈslaɪsɪŋ/ **v.** 切成片，切开（slice 的现在分词）
shaving /ˈʃeɪvɪŋ/ **v.** 刨花（shave 的现在分词）

中文

Jackson：史蒂夫，你需要知道如何处理根茎类蔬菜。有一些常见的根茎类蔬菜，例如洋葱、胡萝卜、白萝卜、土豆、莲藕和南瓜。

Steve：首先，我认为我们应该正确地清理或冲洗它们。然后……切开。

Jackson：正确。每次使用前别忘了磨刀。用磨刀棒保持刀的锋利。在厨房用钝刀工作很危险。

Steve: 我明白，厨师长。然后我们开始切。
Jackson: 在砧板下面放一块湿毛巾，以防止砧板摇摆或滑落。切片和刨花时也需要使用削皮刀。
Steve: 好的，厨师长。我还有很多东西要学。

注释

（1）Knife 翻译为"刀"，包括餐刀、厨房的刀具或者其他刀具。

黄油刀　butter knife
牛排刀　steak knife
水果刀　fruit knife
雕刻刀　carving knife
折刀　pen-knife

注：剪刀　scissors /ˈsɪzəz/

（2）磨刀使用的工具。

磨刀石　sharpening stone　whetstone
磨刀棒　sharpening steel

（3）"砧板"的词汇。

cutting board　chopping board

（4）"peeler"这一词还可以用于：

水果削皮刀　fruit peeler
苹果削皮刀　apple peeler

3. Oven-roasted Chicken 用烤箱做的烤鸡

（初级厨师：Steve，烧腊厨师长：Ben）

Ben: Learning to use an oven with ease is a must. Come and watch, Steve.
Steve: Yes, chef.
Ben: The chicken is ready to go into the oven. Make sure the foil is folded around the roasting tray.
Steve: Yes, it's folded tightly.
Ben: That helps to keep the chicken moist and juicy.
Steve: I see.

词汇

oven /ˈʌvn/ *n.* 烤箱
roast /rəʊst/ *v. adj.* 烤，烤的
with ease /wɪð iːz/ 熟练地，不费力地
must /mʌst/ *n.* 绝对必要的事物
ready /ˈredi/ *adj.* 准备好
foil /fɔɪl/ *n.* 锡纸，锡箔
folded /ˈfəʊldɪd/ *adj.* 折叠的
around /əˈraʊnd/ *pre.* 围绕
roasting tray /ˈrəʊstɪŋ treɪ/ 烤盘
tray /treɪ/ *n.* 托盘
tightly /ˈtaɪtli/ *adv.* 紧紧地，牢固地
moist /mɔɪst/ *adj.* 湿润的
juicy /ˈdʒuːsi/ *n.* 多汁的

Ben: The oven is preheated to 180 degrees Celsius. Now put the roasting tray into the oven. Cook for 2 hours at 180 degrees Celsius. Then raise the temperature to 220℃ and cook for a further 20 minutes. Once the chicken is cooked, transfer to a plate.
Steve: I have to take a note!

preheated /ˌpriːˈhiːtɪd/ *v.* 预先加热，预热
degree /dɪˈɡriː/ *n.* 度数
Celsius /ˈselsiəs/ *adj.* 摄氏的
Fahrenheit /ˈfærənhaɪt/ *adj.* 华氏的
raise /reɪz/ *v.* 提高，升起
temperature /ˈtemprətʃə(r)/ *n.* 温度
further /ˈfɜːðə(r)/ *adj.* 更多的，进一步的
once conj. /wʌns/ 一旦
transfer /trænsˈfɜː(r)/ *v.* 转移
plate /pleɪt/ *n.* 碟子
take a note /teɪk ə nəʊt/ 做笔记，记录

中文

Ben： 一定要学会熟练使用烤箱。斯蒂夫，你过来看看。
Steve： 好的。厨师长。
Ben： 鸡肉准备好进烤箱了。确保锡纸包住整个烤盘了。
Steve： 是的，包得很紧。
Ben： 这有助于保持鸡肉湿润多汁。
Steve： 我明白了。
Ben： 烤箱已经预热到180 摄氏度了。现在我们把烤盘放进烤箱。先在180 摄氏度下烤2 小时。然后把温度提高到220 摄氏度，再烤20 分钟。鸡肉烤熟后，转移到盘子里。
Steve： 我得做笔记才行！

注释

（1）"一定、必须"有以下常见的表达。
名词：must
Learning to use an oven with ease is a must. 一定要学会熟练使用烤箱。
情态助动词：must
You must learn how to use an oven with ease. 你必须学会如何轻松地使用烤箱。
形容词：necessary
Learning to use an oven with ease is necessary. 学会轻松使用烤箱是必要的。
情态动词：have to
You have to learn how to use an oven with ease. 你必须学会如何轻松地使用烤箱。

（2）温度的表达。

摄氏度：degrees Celsius

37℃：37 degrees Celsius

华氏度：degrees Fahrenheit

98.6 ℉：98.6 degrees Fahrenheit

注：华氏度 = 32 + 摄氏度 x1.8。冰的熔点为 32 ℉。

（3）常见的锅具。

pot /pɒt/	锅
pan /pæn/	平底锅
skillet /ˈskɪlɪt/	长柄平底煎锅
steamer /ˈstiːmə(r)/	蒸锅
stewpan /stjuːpæn/	炖锅
stockpot /ˈstɒkpɒt/	汤锅
saucepan /ˈsɔːspən/	煮锅，长柄锅
wok /wɒk/	炒菜锅
pressure cooker /ˈpreʃə kʊkə(r)/	高压锅
rice cooker /raɪs ˈkʊkə(r)/	电饭锅
bain marie /ˌbæn məˈriː/	隔水蒸锅

（4）常见的炉具。

stove /stəʊv/	炉灶
oven /ˈʌvn/	烤箱
gas burner /gæs ˈbəːnə/	煤气炉
grill /grɪl/	烤炉
microwave oven /ˈmaɪkrəweɪv ˈʌvn/	微波炉
induction cooker /ɪnˈdʌkʃn ˈkʊkə(r)/	电磁炉

（5）常见的厨房电器设备。

refrigerator /rɪˈfrɪdʒəreɪtə(r)/	冰箱
blender /ˈblendə(r)/	搅拌器
ventilator /ˈventɪleɪtə(r)/	吸油烟机
slicer /ˈslaɪsə(r)/	切片机
grinder /ˈgraɪndə(r)/	研磨机
juicer /ˈdʒuːsə(r)/	榨汁机

练习：熟悉厨房常见工具词汇

写出下列工具的中文及英文名称。

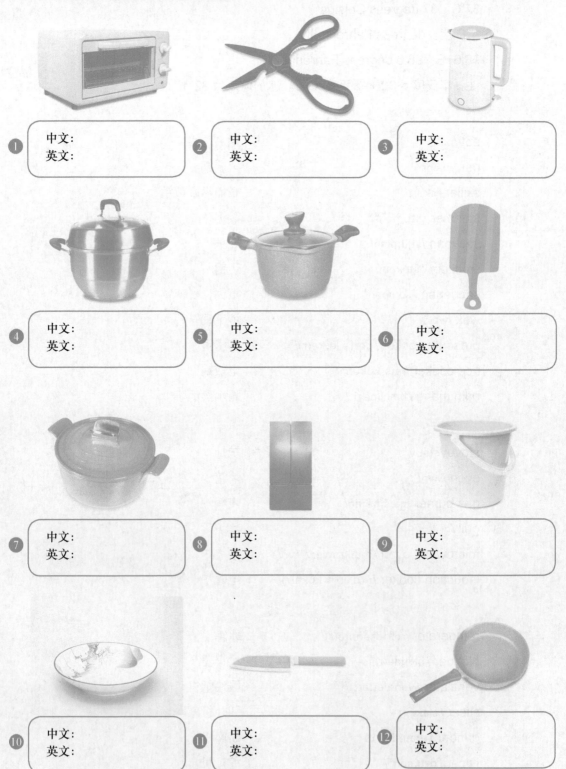

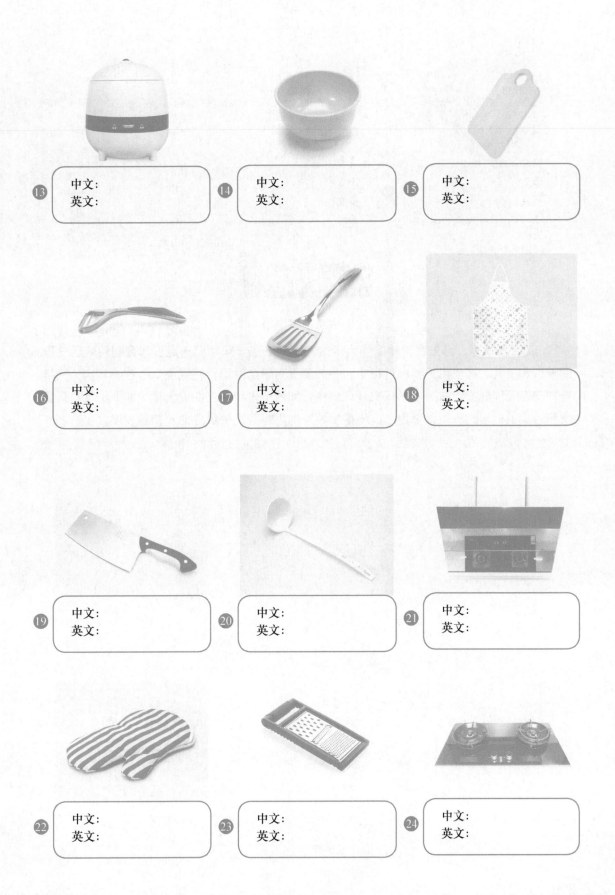

㉕ 中文：
英文：

㉖ 中文：
英文：

职业提示

敬畏生命，遵守规则

"工欲善其事，必先利其器。"为了做好厨师工作，同学们一定要学会利用好厨房里的设备和工具。在使用工具的过程中，首先要提高安全意识，杜绝侥幸心理。其次，严格遵守安全管理制度，按照技术标准规范操作，加强个人和他人的安全保护能力。食物从厨房到舌尖的处理过程中，要用"高质量发展"的思想实现餐饮业的可持续发展。

Part 2
Breakfast
早　　餐

Unit 5　Western Breakfast 西式早餐／044

Unit 6　Chinese Breakfast 中式早餐／054

Unit 7　Fruit Juices and Desserts 果汁和甜点／063

Unit 5　Western Breakfast
西式早餐

提起西式早餐我们就会想到面包（bread）、吐司（toast）、牛角包（croissant）、橙汁（orange juice）、牛奶（milk）等早已融入我们普通家庭餐桌的食品，而提起英式早餐我们也会想到香煎火腿、鸡蛋、肉肠等。酒店及西餐厅主要提供两种主流的西式早餐：欧陆式早餐和英式早餐，食物种类包括面包类、蛋类。

Learning Objectives 学习目标

❉ Memorize the basic terms of Western breakfast.
  记住西式早餐的基本单词和短语。

❉ Know the useful expressions and functional sentences while cooking Western breakfast.
  了解关于制作西式早餐的实用表达和句子。

扫码看视频

Basic Terms 基础词汇

煎蛋　　　　　　　一面煎　　　　　　　两面煎

/fraɪd eg/　　　/ˈsʌni saɪd ʌp/　　　/tɜːnˈəʊvə(r)/
fried egg　　**sunny-side up**　　**turn over**

水煮蛋　　　　　班尼迪克蛋　　　　　炒蛋
　　　　　　　　火腿鸡蛋松饼

/bɔɪld eg/　　　/egz ˈbenədɪkt/　　　/ˈskræmbld eg/
boiled egg　**eggs benedict**　**scrambled egg**

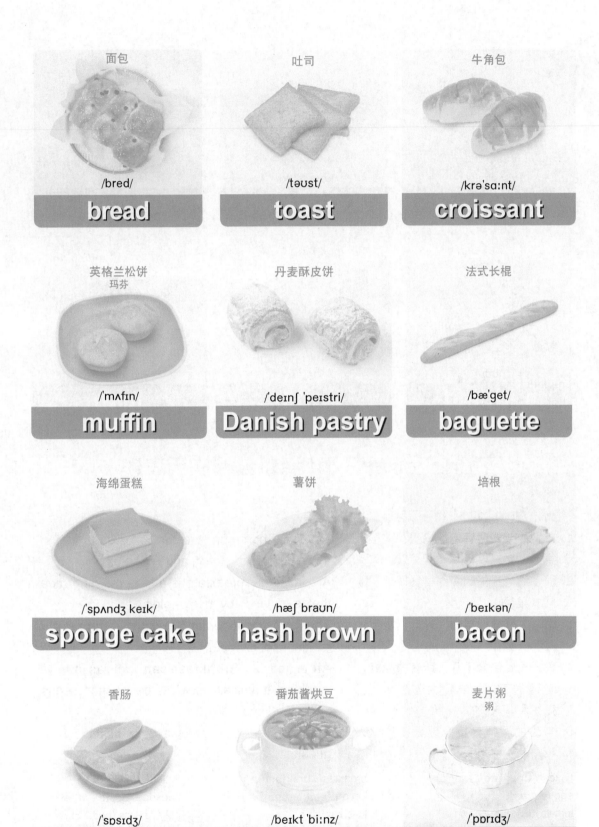

Functional Expressions 实用表达

请流利地朗读以下句子，并做到中英互译。

1. 典型的英式早餐包括鸡蛋、培根、香肠、番茄、吐司和焗豆。
   A typical English breakfast includes eggs, bacon, sausage, tomato, toast, and baked beans.

2. 很多酒店提供的简约早餐便属于欧陆式早餐。
   A continental breakfast is a light morning meal that is provided by many hotels.

3. 术语"欧陆式早餐"来自"欧洲大陆"一词。
   The term "Continental Breakfast" refers to the continent of Europe.

4. 美式早餐包含了几乎所有的元素，它通常包括培根、火腿、鸡蛋、煎饼和华夫饼。
   An American breakfast can include almost anything. It typically includes bacon, ham, eggs, pancakes, and waffles.

5. 当班的厨师直接在现场接受客人下单。
   The on-site chef takes orders directly from guests.

typical /ˈtɪpɪkl/ *adj.* 典型的
egg /eg/ *n.* 蛋
tomato /təˈmɑːtəʊ/ *n.* 西红柿
continental /ˌkɒntɪˈnentl/ *adj.* 大陆的
term /tɜːm/ *n.* 术语
American /əˈmerɪkən/ *adj.* 美国的
breakfast /ˈbrekfəst/ *n.* 早餐
bacon /ˈbeɪkən/ *n.* 培根，熏肉
toast /təʊst/ *n.* 吐司
light /laɪt/ *adj.* 少量的
continent /ˈkɒntɪnənt/ *n.* 大陆，欧洲
almost /ˈɔːlməʊst/ *adv.* 几乎
include /ɪnˈkluːd/ *v.* 包含
sausage /ˈsɒsɪdʒ/ *n.* 香肠
baked beans /beɪkt ˈbiːnz/ 焗豆
provide /prəˈvaɪd/ *v.* 提供
Europe /ˈjʊərəp/ *n.* 欧洲
ham /hæm/ *n.* 火腿

pancake /ˈpænkeɪk/ **n.** 煎饼　　　　waffle /ˈwɒfl/ **n.** 华夫饼　　　　on-site /ˌɒnˈsaɪt/ ***adj.*** 在场的
take order /teɪk ɔːdə(r)/ 点菜，点单　　directly /dəˈrektli/ **adv.** 直接地

6. 我们应该学会如何与客人互动。　　　We should know how to interact with guests.
7. 用"早上好，先生/女士"来问候　　　Greet the guest by saying "Good morning sir/
　　客人。　　　　　　　　　　　　　madam".
8. 蛋类是现做的。　　　　　　　　　　Egg dishes are made-to-order.
9. 你的煎蛋卷要什么馅料？　　　　　　What omelet fillings would you like?
10. 请问你要煎蛋、炒蛋还是水煮蛋？　　Would you like fried egg, scrambled egg, or
　　　　　　　　　　　　　　　　　　boiled egg?

guests /gests/ **n.** 客人（guest 复数）　interact /ˌɪntərˈækt/ **v.** 交流，沟通　greet /griːt/ **v.** 问候
made-to-order /ˌmeɪd tu ˈɔːdə(r)/ 现做的　omelet /ˈɒmlət/ **n.** 煎蛋卷　　　filling /ˈfɪlɪŋ/ **n.** 馅料，填料
fried /fraɪd/ **v.** 煎的　　　　　　　　scrambled /ˈskræmbld/ **v.** 炒（蛋）　boiled /bɔɪld/ ***adj.*** 水煮的

11. 你的鸡蛋要怎么煎？　　　　　　　　How would you like your egg fried?
12. 一面煎还是两面煎？　　　　　　　　Sunny-side up or turn over?
13. 煎蛋是要煎得嫩一些还是全熟？　　　For your fried egg, over easy or over hard?
14. 炒蛋是要嫩一些还是半熟？　　　　　For your scrambled egg, easy scrambled or
　　　　　　　　　　　　　　　　　　medium scrambled?
15. 水煮蛋是要溏心蛋还是全熟蛋？　　　For your boiled egg, soft boiled or hard boiled?

sunny-side up /ˈsʌni saɪd ʌp/ 一面煎，太阳蛋　　　　　turn over /tɜːn ˈəʊvə(r)/ 两面煎（蛋）
easy /ˈiːzi/ ***adj.*** 煮得嫩的（鸡蛋），包括煎蛋、荷包蛋、煎蛋卷　hard /hɑːd/ ***adj.*** 全熟的（鸡蛋）
medium /ˈmiːdiəm/ ***adj.*** 半熟的（鸡蛋）　　　　　soft /sɒft/ ***adj.*** 煮得嫩的（水煮蛋）

16. 荷包蛋是要温泉蛋还是全熟蛋？　　　For your poached egg, soft poached or hard poached?
17. 吐司是轻微烤一下还是烤焦一些？　　For your toast, light or dark?
18. 这是薄烤饼的糖浆，请自便。　　　　Here are the pancake syrups. Help yourself.
19. 需要帮您把碟子端过去吗？　　　　　May I help you with the plate?
20. 请拿好您的碗。　　　　　　　　　　Please hold your bowl steady.

poached /pəʊtʃt/ ***adj.*** 水煮的（低温水煮）　light /laɪt/ ***adj.*** 颜色浅的　　　dark /dɑːk/ ***adj.*** 颜色深的
syrup /ˈsɪrəp/ **n.** 糖浆　　　　　　　　　　　plate /pleɪt/ **n.** 碟子　　　　hold /həʊld/ **v.** 握住
bowl /bəʊl/ **n.** 碗　　　　　　　　　　　　　steady /ˈstedi/ ***adj.*** 牢固的

Task 任务　　情景对话

试着用英文点一个煎蛋：两面煎，嫩一些。

地点：Open Kitchen

人物：初级厨师 Steve，客人 Jack

Dialogues 对话

（初级厨师：Steve，客人：Jack）

1. At the Egg Station 在煎蛋档

Steve: We serve make-to-order fried eggs, boiled eggs, poached eggs, scrambled eggs, and omelets. Which would you prefer?
Jack: Well, one fried egg, sunny-side up. And one omelet as well, please.
Steve: What omelet fillings would you like?
Jack: Uh, cheese, tomatoes, and ham on the top.

词汇

serve /sɜːv/ v. 招待，为……服务
fried eggs /fraɪd egz/ 煎蛋
boiled eggs /bɔɪld egz/ 水煮蛋
poached eggs /pəʊtʃt egz/ 荷包蛋
scrambled eggs /ˈskræmbld egz/ 炒蛋
omelet /ˈɒmlət/ n. 煎蛋卷
prefer /prɪˈfɜːr/ v. 更喜欢
sunny-side up /ˈsʌni saɪd ʌp/ 一面煎
as well /æz wel/ 也，还有，此外
omelet filling /ˈɒmlət fɪlɪŋ/ 煎蛋卷馅料
cheese /tʃiːz/ n. 奶酪，芝士
ham /hæm/ n. 火腿
on the top /ɒn ðə tɒp/ 在上面，在顶部

中文

Steve：我们有现做的煎蛋、水煮蛋、荷包蛋、炒蛋和煎蛋卷。您想要哪种？
Jack：嗯，一个煎蛋，一面煎。还要一个煎蛋卷。
Steve：您的煎蛋卷要放些什么馅料？
Jack：呃，上面放芝士、西红柿和火腿。

注释

（1）"well"做语气词，在不同的语境下可以翻译为不同的中文。
Well, we'd better go home. 呃，我们该回家了。
Well, really? 啊……真的吗？
Well, just do it. 好吧，去做就是了。
Well, I was just trying to explain. 噢，我只是试着解释。
（2）常见的语气词。

My God! / God! 天啊（惊叹）
Look out! 小心（警告）
Wow /waʊ/ 哇（惊喜）
Hooray /huˈreɪ/ 万岁（好极了）
Uh-huh /ʌˈhʌ/ 嗯（同意）
Oui /wiː/ 是，对（同意）
Yuck /jʌk/ 恶心（厌恶）

Ouch /aʊtʃ/ 哎哟（表示突然的疼痛）
Uh/Er/Eh /ʌ; ɜː/ 哦，嗯（表示回应）
Oops /uːps/ 哎呀（差点出事故、摔坏物品时）
Ah /ɑː/ 啊（羡慕）
Ha-ha 哈哈（笑）
Aye /aɪ/ 哎（痛苦或吃惊）
Eww /ˈiːuː/ 喷（嫌弃）

> （3）"make-to-order" 理解为：根据订单而制作；定制；现做现卖。还有以下常见表达：
>
> 现磨咖啡 freshly brewed coffee
> 鲜榨果汁 freshly squeezed juice
> 现成的点心 ready-made pastry

2. Making an Eggs Benedict 做一份班尼迪克蛋

（初级厨师：Steve，行政副厨师长：Jackson）

Jackson:	Now learn how to make an eggs benedict step by step, Steve. Step 1, use English muffins as the base of eggs benedict.
Steve:	Step 2, make hollandaise sauce. I got this.
Jackson:	That's right. Step 3, cook 2 poached eggs. The eggs are supposed to be runny.
Steve:	Absolutely, chef.
Jackson:	A slice of bacon comes next. Don't get too stuck on the bacon. Ham and smoked salmon are popular options as well.
Steve:	Oui, chef.
Jackson:	At last, top it off with hollandaise sauce. It's done.
Steve:	Serve immediately. Delicious! (Grinning)

词汇

eggs benedict /ˌegz ˈbenədɪkt/ 班尼迪克蛋，火腿蛋松饼
step by step /step baɪ step/ 逐步地，一步一步来
muffin /ˈmʌfɪn/ n. 松饼，玛芬
base /beɪs/ n. 基底，基础
hollandaise /ˌhɒləndeɪz/ 荷兰酱（由蛋黄、黄油、醋或柠檬汁混合而成）
sauce /sɔːs/ n. 酱汁，沙司
supposed /səˈpəʊzd/ adj. 应该（表示按计划或期望）
runny /ˈrʌni/ adj. 水分过多的，软的
absolutely /ˈæbsəluːtli/ adv. 绝对的，当然
get stuck on /get stʌk ɒn/ 被卡住，被困住
smoked salmon /sməʊkt ˈsæmən/ 烟熏三文鱼
option /ˈɒpʃn/ n. 选项
oui /wiː/ int. 是，对
at last /æt læst/ 最后
top off /tɒp ɒf/ 完成，结束（这里也指从顶部淋下荷兰酱）
serve /sɜːv/ v. 招待，为……服务
immediately /ɪˈmiːdiətli/ adv. 立即地
delicious /dɪˈlɪʃəs/ adj. 美味的
grinning /ˈɡrɪnɪŋ/ n. 露齿而笑，咧嘴笑

中文

Jackson：现在来学习制作班尼迪克蛋的步骤，斯蒂夫。第一步，用英式松饼来给班尼迪克蛋打底。
Steve：第二步，做荷兰酱。这个我会做。
Jackson：没错。第三步，煮两个荷包蛋。这些蛋最好是溏心蛋。
Steve：当然，厨师长。

Jackson： 接下来是培根片。不要太执着于用培根。火腿和烟熏三文鱼也是受欢迎的选项。
Steve： 是的，厨师长。
Jackson： 最后，加荷兰酱调味。完成了。
Steve： 即可食用。美味！（露齿笑）

注释

（1）"eggs benedict"是美式早餐里最负盛名的一道菜。一份标准的班尼迪克蛋中含有加拿大培根或火腿片、荷包蛋、荷兰酱以及英式松饼。在说到早午餐（brunch）时，许多人脑海里浮现出的第一道菜就是班尼迪克蛋。

（2）蛋黄处于还未凝固、流动的状态时，英文使用"runny"来形容。在本对话里则直接翻译为"溏心蛋"。蛋黄的英文词汇是"yolk"，那么不同的溏心蛋还可以这样表达：

boiled runny yolk egg　　水煮溏心蛋
poached runny yolk egg　　荷包溏心蛋
fried runny yolk egg　　煎溏心蛋

3. The Pancakes, Waffles and Toppings 煎饼、华夫饼和配料

（初级厨师：Steve，行政副厨师长：Jackson）

Jackson: Pancakes and waffles are popular breakfast foods. The ingredients of pancake and waffle batter are almost the same.
Steve: I thought they share a same batter.
Jackson: Nope. Besides, the pancake is cooked in small rounds of a frying pan or griddle. Waffles are cooked in a waffle iron.

词汇

pancake /'pænkeɪk/ n. 煎饼，烤薄饼
waffle /'wɒfl/ n. 华夫饼
ingredient /ɪn'griːdiənt/ n. 原料
batter /'bætə(r)/ n. 面糊，糊状物
thought /θɔːt/ v. 认为（think 的过去式和过去分词）
share /ʃeə(r)/ v. 分享，共用
nope /nəʊp/ adv. 没有；不
besides /bɪ'saɪdz/ adv. 此外；以及；也
rounds /raʊndz/ n. 圆型物
frying pan /'fraɪɪŋ pæn/ 煎锅，长柄平底锅
griddle /'grɪdl/ n. 煎饼用浅锅
iron /'aɪən/ n. 铁做的工具

Steve:	I see. The method of cooking is different.
Jackson:	Yes. And there are many toppings that go well with them. Maple syrup is a classic pancake and waffle topping. We offer our guests chocolate syrup, cream, honey, and so on.
Steve:	Wow, that sounds great!

method /ˈmeθəd/ **n.** 方法
different /ˈdɪfrənt/ **adj.** 不同的
topping /ˈtɒpɪŋ/ **n.** （菜肴、蛋糕等上的）浇汁，浇料，配料，佐料
go with /gəʊ wɪð/ 搭配
go well with /gəʊ wel wɪð/ 协调，和……很相配
maple syrup /ˈmeɪpl ˈsɪrəp/ 枫糖浆
classic /ˈklæsɪk/ **adj.** 经典的
chocolate syrup /ˈtʃɒklət ˈsɪrəp/ 巧克力糖浆
cream /kriːm/ **n.** 奶油
honey /ˈhʌni/ **n.** 蜂蜜

中文

Jackson：	薄煎饼和华夫饼是受欢迎的早餐食品。煎饼和华夫饼的面糊成分几乎相同。
Steve：	我以为它们用同一种面糊。
Jackson：	不对。此外，煎饼是用有小圆孔的平底锅或铁盘做的。华夫饼是用华夫饼烘烤模来做的。
Steve：	我明白了。烹饪方法不同。
Jackson：	是的。还有很多可以搭配淋在上面的配料。枫糖浆是经典的煎饼和华夫饼的淋酱。我们为客人提供巧克力糖浆、奶油和蜂蜜等。
Steve：	哇，听起来很棒！

注释

（1）煎饼（美式）和华夫饼差异对比。

煎饼： 1）面粉、发酵粉、糖、鸡蛋、黄油、盐和牛奶。
Pancake 2）所有成分一次性混合。
　　　　3）使用平底锅或铁盘。

华夫饼： 1）面粉、发酵粉、糖、鸡蛋、黄油、盐、牛奶和香草精。
Waffle 2）混合过程分几步进行，蛋清搅打均匀后再加入面糊。
　　　　3）使用华夫饼专用模具。

（2）在美国，"pancake"有时也叫作"hotcake"。

（3）当我们在描述菜肴的顶部配料时，可以使用英文词汇"topping"，或者需要表达"在……顶上放"的意思时，使用英文词组"top...with"。比如：

华夫饼顶部的配料 / 装饰物	waffle toppings
华夫饼上加冰淇淋	waffles topped with ice cream
在华夫饼上放点蓝莓	top the waffle with blueberries
在奶油上点缀一些草莓粒	top the cream with chopped strawberries

> *Just choose some waffle toppings that fit your liking.
> 只要选些你喜欢的华夫饼配料就行了。

练习：掌握煎蛋的常见表达

西式早餐会有各种鸡蛋菜式 (egg dishes)，比如一面煎的太阳蛋和两面煎的蛋，炒鸡蛋和搭配了不同配料的煎蛋卷，还有更复杂一些的班尼迪克蛋。大部分情况下这些菜式都是现点现做的，厨师需要在档口与客人面对面沟通。完成以下练习：

1. 翻译词汇 Translate basic vocabularies.

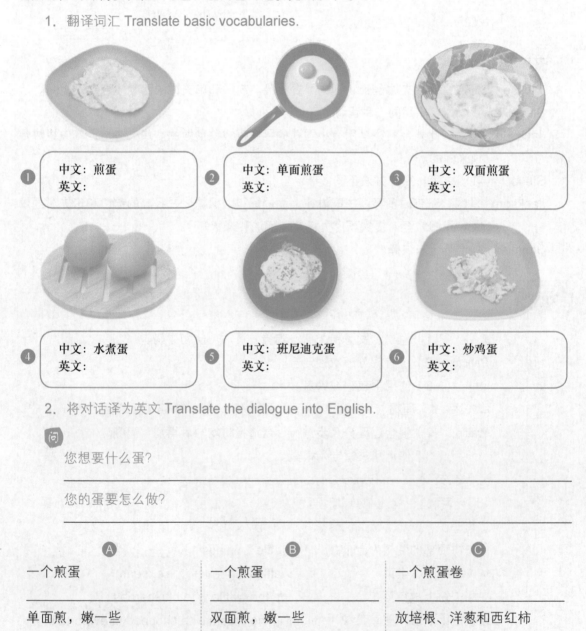

① 中文：煎蛋　英文：
② 中文：单面煎蛋　英文：
③ 中文：双面煎蛋　英文：
④ 中文：水煮蛋　英文：
⑤ 中文：班尼迪克蛋　英文：
⑥ 中文：炒鸡蛋　英文：

2. 将对话译为英文 Translate the dialogue into English.

您想要什么蛋？
_____

您的蛋要怎么做？
_____

Ⓐ	Ⓑ	Ⓒ
一个煎蛋	一个煎蛋	一个煎蛋卷
单面煎，嫩一些	双面煎，嫩一些	放培根、洋葱和西红柿

D 一个水煮蛋
_____

E 一个溏心水煮蛋
_____

F 一份炒蛋
_____

职业提示

合理早餐，健康为先

"一日之际在于晨。"早餐是我们每天充足活力的源泉。作为厨师，我们需要科学、合理地为客人设计富有营养的早餐食谱，包括丰富的碳水化合物、优质的蛋白质及高纤维蔬果。早餐品种应该花样丰富、主副搭配、荤素互补，满足不同客人的需求。在党的二十大报告中，强调了"推进健康中国建设，深入实施健康中国战略"，提出了"全民健康意识日益增强，健康服务体系逐步完善，健康产业加快发展"的目标和要求。我们相信，只有将健康意识融入餐饮服务的方方面面，才能更好地服务于国家和人民的健康事业。

Unit 6 Chinese Breakfast
中式早餐

总结起来，西式早餐多为吐司、麦片、煎饼、火腿、三明治、鸡蛋等。而中国是一个幅员辽阔、人口众多、文化迥异的国家，不同地区的情况也不尽相同。早餐包含多种不同的口味和菜肴。中式早餐必须是热的，准备速度要快。南方的早餐相比北方的早餐口味清淡，辅以各种辣椒、小菜。北方的早餐口味较重，以咸最多见，油重、口感扎实，多放葱蒜，以面食为主。

Learning Objectives 学习目标

✵ Memorize the basic terms of Chinese breakfast.
  记住中式早餐的基本单词和短语。

✵ Know the useful expressions and functional sentences while cooking Chinese breakfast.
  了解关于制作中式早餐的实用表达和句子。

扫码看视频

Basic Terms 基础词汇

Unit 6　Chinese Breakfast 中式早餐 /055

包子 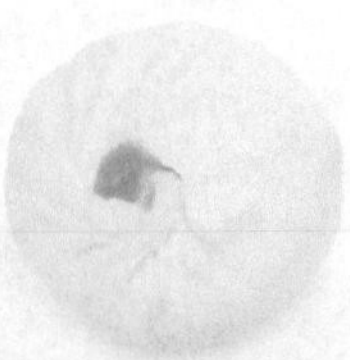
baozi
Steamed bun stuffed with (filling)

叉烧包 
Char Siu Bao
steamed bun stuffed with barbecued Pork

饺子 
Jiaozi

馒头
mantou
steamed bread

粽子 
zongzi
glutinous rice wrapped in bamboo leaves

广式点心 
/ˌdɪm ˈsʌm/
dim sum

糯米 
/ˈgluːtənəs raɪs/
glutinous rice

汤圆 
tangyuan
glutinous rice ball

芝麻汤圆 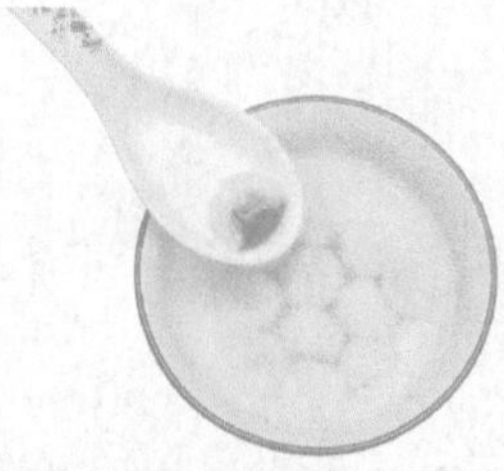
glutinous rice balls stuffed with sesame

一碗粥 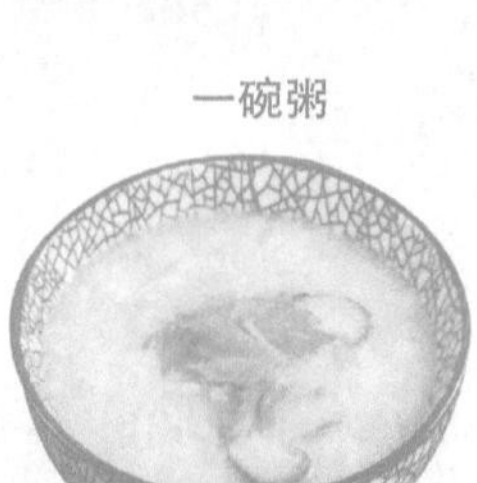
a bowl of congee
a bowl of rice porridge

鱼片粥 
sliced fish congee
fish fillet congee

皮蛋瘦肉粥 
congee with pork and preserved egg

炒面	牛肉面	桂林米粉
fried noodles	beef noodles	Guilin rice noodles
豆腐花	泡菜，咸菜 /ˈpɪkl/	烧饼
tofu pudding	pickle	Chinese pancake

Functional Expressions 实用表达

请流利地朗读以下词组，并做到中英互译。

1. 一碟小笼包	a plate of baozi
2. 叉烧包	baozi stuffed with BBQ Pork
3. 这个包子里面是什么馅料？	what's the stuffing/filling in this bun?
4. 一碗牛肉面	a bowl of beef noodles
5. 一碗米粉	a bowl of rice noodles

plate /pleɪt/ **n.** 碟
BBQ = barbeque /ˈbɑːbəkjuː/ 烧烤
filling /ˈfɪlɪŋ/ **n.** 填充物

stuffed /stʌft/ **adj.** 酿，填充的
stuffing /ˈstʌfɪŋ/ **n.** 填充，材料
bowl /bəʊl/ **n.** 碗

6. 螺蛳粉	luosifen (Liuzhou river snail rice noodles)
7. 一碗粥	a bowl of congee
8. 皮蛋瘦肉粥	congee with pork and preserved egg
9. 一杯豆浆	a cup of soybean milk
10. 粽子	zongzi (glutinous rice wrapped in bamboo or lotus leaves)

river snail /ˈrɪvə(r) sneɪl/ 田螺
pork /pɔːk/ **n**. 猪肉
cup /kʌp/ **n**. 杯，杯子
soybean milk /ˈsɔɪbiːn mɪlk/ 豆浆
wrapped /ræpt/ **v**. 包裹，覆盖
lotus leaf /ˈləʊtəs liːf/ 荷叶

congee /ˈkɒndʒiː/ **n**. 粥
preserved egg /prɪˈzɜːvd eg/ 皮蛋
soybean /ˈsɔɪbiːn/ 大豆
glutinous rice /ˈgluːtənəs raɪs/ 糯米
bamboo leaf /ˌbæmˈbuː liːf/ 竹叶
leaves /liːvz/ **n**. 树叶，叶片，花瓣（leaf 的复数）

11. 芝麻汤圆	glutinous rice balls stuffed with sesame
12. 油条	youtiao (deep-fried dough sticks)
13. 豆腐花	tofu pudding
14. 韭菜猪肉饺子	jiaozi stuffed with Chinese leek and pork
15. 榨菜	pickled Chinese mustard tuber

ball /bɔːl/ **n**. 球
dough /dəʊ/ **n**. 生面团
pudding /ˈpʊdɪŋ/ **n**. 布丁
pickled /ˈpɪkld/ **adj**. 腌制的；盐渍的

sesame /ˈsesəmi/ **n**. 芝麻
stick /stɪk/ **n**. 棍子
jiaozi 饺子
mustard /ˈmʌstəd/ **n**. 芥菜，芥末

deep-fried /ˌdiːpˈfraɪd/ **adj**. 炸的
tofu /ˈtəʊfuː/ **n**. 豆腐
Chinese leek /tʃaɪniːz liːk/ 韭菜
tuber /ˈtjuːbə(r)/ **n**. 块茎

16. 在粥里加点咸菜。	get a bit of pickles with congee.
17. 我们喜欢蘸着酱油和醋吃饺子。	we love to dip jiaozi in soy and vinegar sauce.
18. 广式点心	Cantonese dim sum
19. 这些各种各样的小菜叫点心。	These various small dishes are called dim sum.
20. 我们在摊贩处买早餐。	We take breakfast from the street food vendors.

pickle /ˈpɪkl/ **n**. 泡菜；盐卤；腌制食品
soy /sɔɪ/ **n**. 酱油
sauce /sɔːs/ **n**. 酱汁，沙司
dim sum /ˌdɪm ˈsʌm/ 点心（粤语音译）
vendor /ˈvendə(r)/ **n**. 卖主，小贩

dip /dɪp/ **v**. 浸，蘸
vinegar /ˈvɪnɪgə(r)/ **n**. 醋
Cantonese /ˌkæntəˈniːz/ **adj**. 广东人的 **n**. 广东话
various /ˈveəriəs/ **adj**. 各种各样的
street vendor /striːt ˈvendə(r)/ 街贩

Task 任务　情景对话

请设计一段简单的对话，告诉站档的厨师你想要什么中式早餐。

地点：Diamond Restaurant

人物：客人 Mr.White，初级厨师 Steve

💬 Dialogues 对话

（初级厨师：Steve，客人：Bella，副厨师长：Jackson）

1. At the Congee Station 在粥档

（初级厨师：Steve，中餐厨师长：David）

David: Congee is a classic Chinese breakfast dish. It's very good for babies, kids, and persons who feel under the weather.

Steve: My mom always feed me congee when I was little.

David: It's an important staple on the Chinese dinner table. Cooking plain congee is not as easy as it looks.

Steve: I understand, Chef.

David: Mix the medium grain and long grain rice. Soak the rice in water overnight.

Steve: I see. Can we use the instant pot?

David: I prefer the clay pot. The rice-water ratio is about 1:20. Bring the rice to a boil over high heat. Then simmer for 2 hours.

Steve: And we have to stir the congee a few times during cooking. It really is not as easy as it looks.

词汇

Chinese Chef 中餐厨师长
classic /ˈklæsɪk/ adj. 经典的
babies /ˈbeɪbɪz/ n. 婴儿（baby 的复数）
kids /kɪdz/ n. 小孩（kid 的复数）
feel under the weather /fiːl ˈʌndə(r) ðə ˈweðə(r)/ 不舒服，有点小病
feed /fiːd/ v. 喂食，喂养
staple /ˈsteɪpl/ n. 主食
plain congee /pleɪn kɒndʒiː/ 白粥
mix /mɪks/ v. 混合
medium grain /ˈmiːdiəm ɡreɪn/ 中等颗粒
grain /ɡreɪn/ n. 颗粒；谷物
soak /səʊk/ v. 浸泡，渗透
overnight /ˌəʊvəˈnaɪt/ adv. 一夜之间，过夜地
instant pot /ˈɪnstənt pɒt/ 电压力锅
instant /ˈɪnstənt/ adj. 立即的，即时的
clay pot /kleɪ pɒt/ 瓦煲，黏土锅
ratio /ˈreɪʃiəʊ/ n. 比例
1:20 读作 one to twenty
bring to a boil /brɪŋ tu ə bɔɪl/ 煮至沸腾
high heat /haɪ hiːt/ 大火，高温
simmer /ˈsɪmə(r)/ v. 炖，煨
stir /stɜː(r)/ v. 搅拌

中文

David：粥是经典的中式早餐。对婴儿、小孩和不舒服的人来说非常好。

Steve：小时候我妈妈总是喂我粥。

David：它是中国人餐桌上的重要主食。煮白粥并不像看起来那么容易。

Steve：我明白，厨师长。

David：将中等颗粒的米和长粒的米混合。在水中浸泡一晚。

Steve：好的。我们可以使用高压锅吗？

David：我更愿意用瓦罐煲。米和水的比例约为 1:20，用大火煮沸，然后小火煮 2 小时。

Steve：我们还得在煮的过程中多次搅拌。它真的并不像看起来那样容易。

注释

粥是除了米饭之外中国人餐桌上最常见的主食。依照口味的不同，各家都有自己的煮粥诀窍。下面是常见的粥品及对应的英文翻译：

鱼片粥 fish fillet congee　　猪骨粥 pork bone congee
海鲜粥 seafood congee　　　绿豆粥 mung bean congee

地瓜粥 sweet potato congee	腊八粥 congee with nuts and dried fruits
小米粥 millet congee	白粥 plain congee

2. At the Noodle Station 在面档

（初级厨师：Steve，客人：Bella）

Steve: Good morning, Madam. Would you like some noodles?
Bella: I'd like to try this noodle soup.
Steve: No problem. Here are noodles, rice noodles, and wide rice noodles. Firstly, which staple food would you like?
Bella: Noodles, please.
Steve: Then you choose the side dishes, such as spinach, lettuce, fungus, mushrooms, or meatballs.
Bella: Er...I'll have spinach, fungus, and 2 meatballs, please.
Steve: OK. It will take about 3 minutes. Please come back here when the noodle is cooked.
Bella: I see.
　　　...3 minutes later
Steve: Madam, your dish is ready. Here is the bowl, hold on to it. Be careful, it's hot.
Bella: Thanks. What condiments can I put on for flavoring?
Steve: I would suggest you add some scallions, soy sauce, chilli sauce, and pickles.
Bella: Thanks. Let me see.

词汇

noodles /ˈnuːdlz/ n. 面条（noodle 的复数）
noodle soup /ˈnuːdl suːp/ 汤面，面汤
rice noodles /raɪs ˈnuːdlz/ 米粉
wide rice noodles /waɪd raɪs ˈnuːdlz/ 宽米粉，扁粉
staple /ˈsteɪpl/ n. 主食
choose /tʃuːz/ v. 选择
side dish /ˈsaɪd dɪʃ/ 配菜，小菜，附加菜
spinach /ˈspɪnɪtʃ/ n. 菠菜
lettuce /ˈletɪs/ n. 生菜
fungus /ˈfʌŋɡəs/ n. 菌，菌类，木耳
mushroom /ˈmʌʃrʊm/ n. 蘑菇
meatball /ˈmiːtbɔːl/ n. 肉丸
minute /ˈmɪnɪt/ n. 分钟
cooked /kʊkt/ adj. 煮熟的
dish /dɪʃ/ n. 一碟菜，盘子
ready /ˈredi/ adj. 准备好
bowl /bəʊl/ n. 碗
hold on to it /həʊld ɒn tu ɪt/ 拿好
condiment /ˈkɒndɪmənt/ n. 调味品
flavoring /ˈfleɪvərɪŋ/ n. 调味；v. 加味于……
suggest /səˈdʒest/ v. 建议
scallion /ˈskæliən/ n. 葱
soy sauce /ˌsɔɪ ˈsɔːs/ 酱油
chilli sauce /ˈtʃɪli sɔːs/ 辣椒酱
pickle /ˈpɪkl/ n. 小菜，泡菜，咸菜

中文

Steve：早上好，女士。要来些面条吗？
Bella：我想尝尝这种汤面。
Steve：没问题。这是面条、米粉和宽米粉。首先，您想要哪种主食？
Bella：面条。
Steve：然后您再选择配菜，比如菠菜、生菜、木耳、蘑菇或肉丸。

Bella：嗯，我要菠菜、木耳和两个肉丸吧。
Steve：好的。大概需要三分钟。等到面条煮熟以后再回来这里。
Bella：好。

……三分钟以后

Steve：女士，您的面条好了。这是您的碗，端好。小心烫。
Bella：谢谢。我可以放哪些调味品？
Steve：我建议你加些葱、酱油、辣椒酱和咸菜。
Bella：谢谢。我想想。

> **注释**
>
> 米粉在翻译为英文时借用了"noodle"一词，意为"米做的面条"。而有时候"米粉"也使用音译以及借用其他国家关于面食的词汇进行翻译。比如：
>
> （1）Ho Fun 河粉（一种宽米粉），根据粤语里河粉的发音而采用的英文翻译。类似的还有"Chao Fun"（炒粉）及"Cheung Fun"（肠粉）。
>
> （2）Pho 越南河粉，越南语发音类似 /fʌ/，/fə/；英文发音为 /fəʊ/。
>
> （3）Vermicelli /ˌvɜːmɪˈtʃeli/ *n.* 原指意式细面，在中餐菜单里译为细而长的米粉丝。
>
> 如：fried rice vermicelli 炒米线

3. Dim Sum Cart 点心车

（初级厨师：Steve，客人：Bella）

Steve: Good morning, Madam.
Bella: Excuse me, what's this?
Steve: This is our "dim sum" cart. There is a variety of different dim sum dishes. You can choose whatever you like.
Bella: Do you have a menu?
Steve: Sure. You can order from the cart and a menu.
Bella: Wow, interesting. This is ...
Steve: This is barbecued pork bun, char siu bao. It's a must-try. Take one?
Bella: I'll have this, please. Thank you.
Steve: I will come back later. Be ready when the cart rolls by!

词汇

cart /kɑːt/ *n.* 购物车，推车
variety /vəˈraɪəti/ *n.* 多样；种类
dishes /ˈdɪʃɪz/ *n.* 菜肴（dish 的复数）
choose /tʃuːz/ *v.* 选择
whatever /wɒtˈevə(r)/ *det.* 任何事物；不管什么
menu /ˈmenjuː/ *n.* 菜单
order /ˈɔːdə(r)/ *v.* 订购，点单
interesting /ˈɪntrəstɪŋ/ *adj.* 有趣的
barbecued /ˈbɑːbɪkjuːd/ *adj.* 烧烤的
pork bun /pɔːk bʌn/ 肉包，猪肉包
char siu bao 叉烧包（粤语音译）
must-try /mʌst traɪ/ 一定得尝尝的，不可不尝的

中文

Steve：早上好，女士。
Bella：不好意思，这是什么？
Steve：这是我们的"点心"推车。这里有各种各样的点心。您可以选择任何您喜欢的。
Bella：你有菜单吗？
Steve：当然。您可以从点心车和菜单中点菜。
Bella：哇，有趣。这是……
Steve：这是叉烧包。这是必须尝试的。拿一个？
Bella：请给我。谢谢。
Steve：我待会再回来。点心车经过时请做好准备哦！

注释

20世纪50年代开始，广式点心在美国、英国、加拿大、澳大利亚等地因华人移民潮渐起而开始驰名。我国香港的中餐师傅们把广式点心带出亚洲，收获了许多的中华饮食文化的粉丝。以下是广式点心里常见的小吃及其英文译名：

虾饺	shrimp dumpling
烧麦	siu mai (dumplings with pork and shrimp)
糯米鸡	sticky rice in lotus leaf
蒸排骨	steamed pork ribs
蒸牛百叶	steamed beef tripe
鲜虾肠粉	steamed rice roll with shrimp
蒸凤爪	steamed chicken feet
咸水角	glutinous rice dumplings
香煎萝卜糕	pan fried turnip cake

我们发现有些小吃的民间译名非常有趣，例如广式点心里的叉烧，在许多国外唐人街的菜单里直接叫作"char siu"，因其粤语发音而得名。久而久之，叉烧包的英文翻译就真的是"char siu bao"了。当然，还需要在菜单上备注"steamed bun stuffed with barbecued pork"。类似的译名还有饮茶（yum cha）、馄饨（wonton）、捞面（lo mein）、炒面（chow mein）等。

练习：中式小吃翻译

翻译以下食物，并背诵词汇。

序号	中文	英文	序号	中文	英文	序号	中文	英文
1	米粉		3	八宝粥		5	豆腐花	
2	米		4	糯米		6	榨菜	
7	包子		15	豆浆		23	皮蛋瘦肉粥	
8	饺子		16	烧饼		24	芝麻汤圆	
9	馒头		17	叉烧包		25	油条	
10	甜面酱		18	韭菜猪肉饺子		26	广式点心	
11	面条		19	玉米馒头		27	豆沙包	
12	小米粥		20	辣椒酱		28	馄饨	
13	南瓜粥		21	粥		29	花卷	
14	粽子		22	一碗粥		30	肠粉	

职业提示

遵守规范,注重细节

"色香味形"俱佳的菜肴能让顾客赞不绝口,而烹调过程的标准化是决定美食的关键。在正规的餐饮机构,清洗、称量、配菜、烹制、装盘都有着严格的标准化流程。我们在平时的实操中就要养成标准化操作意识,每一步都按照流程表操作,保证菜品质量。

Unit 7 Fruit Juices and Desserts
果汁和甜点

关于菜单、果汁、糕点、口味、气味、原材料等的英文表达里都会接触到水果的词汇。比如，菜单里有两道菜"Passion Fruit Mousse"和"Chilled Avocado Soup"，在水果词汇都掌握了的情况下，我们很轻易就知道这是百香果慕斯和牛油果冻汤。在开始进入果汁及甜点的学习时，需要额外去记忆水果英文词汇。

Learning Objectives 学习目标

❄ Memorize common fruits and desserts vocabularies.
  记住常见水果和点心的词汇。
❄ Learn some expressions of fruit and dessert dishes on menu.
  学习一些菜单上水果和甜点的表达。

扫码看视频

Basic Terms 基础词汇

苹果	香蕉	梨
/ˈæp(ə)l/	/bəˈnɑːnə/	/peə(r)/
apple	**banana**	**pear**

橘子	桃子	柠檬
/ˌtændʒəˈriːn/	/piːtʃ/	/ˈlemən/
tangerine	**peach**	**lemon**

草莓

/ˈstrɔːbəri/
strawberry

黑莓
/ˈblækbəri/
blackberry

蔓越莓
越橘、小红莓

/ˈkrænbəri/
cranberry

树莓
覆盆子

/ˈrɑːzbəri/
raspberry

杨梅

/ˈwæksbəri/
waxberry
red beyberry

樱桃

/ˈtʃeri/
cherry

木瓜

/pəˈpaɪə/
papaya

菠萝
/ˈpaɪnæpl/
pineapple

榴莲

/ˈdʊəriən/
durian

芒果

/ˈmæŋɡəʊ/
mango

菠萝蜜

/ˈdʒækfruːt/
jackfruit

百香果

/ˈpæʃ(ə)n fruːt/
passion fruit

Unit 7　Fruit Juices and Desserts 果汁和甜点 /065

西瓜

/ˈwɔːtəmelən/
watermelon

蜜瓜

/ˈhʌnɪdjuːˈmelən/
honeydew melon
sweet melon

香瓜

/ˈmʌskˌmelən/
muskmelon

无花果

/fɪɡ/
fig

椰子

/ˈkəʊkənʌt/
coconut

火龙果

/ˈdræɡən fruːt/
dragon fruit
pitaya

牛油果
鳄梨

/ˌævəˈkɑːdəʊ/
avocado

番石榴
芭乐

/ˈɡwɑːvə/
guava

柚子

/ˈpɒmɪləʊ/
pomelo

石榴

/ˈpɒmɪɡrænɪt/
pomegranate

葡萄

/ɡreɪp/
grape

龙眼

/ˈlɒŋɡən/
longan

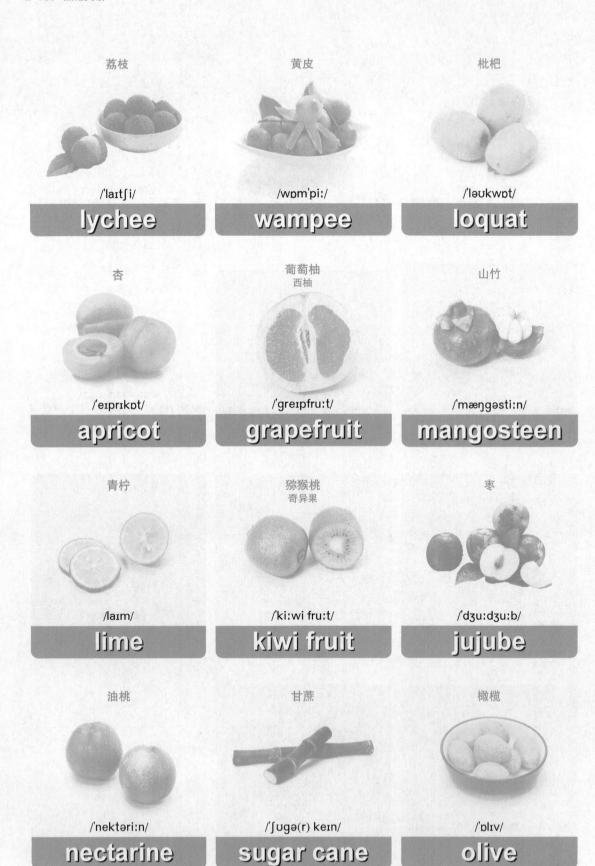

Unit 7　Fruit Juices and Desserts 果汁和甜点 /067

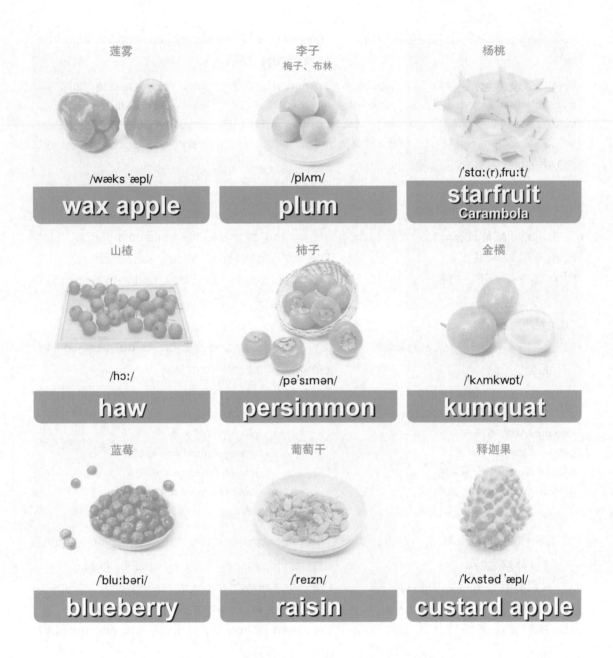

莲雾	李子 梅子、布林	杨桃
/wæks ˈæpl/ **wax apple**	/plʌm/ **plum**	/ˈstɑː(r)ˌfruːt/ **starfruit** Carambola
山楂	柿子	金橘
/hɔː/ **haw**	/pəˈsɪmən/ **persimmon**	/ˈkʌmkwɒt/ **kumquat**
蓝莓	葡萄干	释迦果
/ˈbluːbəri/ **blueberry**	/ˈreɪzn/ **raisin**	/ˈkʌstəd ˈæpl/ **custard apple**

Functional Expression 实用表达

请流利地朗读以下句子，并做到中英互译。

1. 我们有好几种口味的冰淇淋，比如菠萝味、草莓味和香草味。
   We have several ice cream flavors, like pineapple, strawberry, and vanilla.
2. 你喜欢什么口味？
   What flavor would you like?
3. 一旦你尝过榴莲，你不可能会忘记这种经历。
   Once you have tasted durian, it is an experience you are not likely to forget.
4. 挖一勺西瓜味的冰淇淋球。
   Scooping a ball of watermelon ice cream.

5. 红色的是草莓酱，蓝色的是蓝莓酱。 The red one is strawberry jam, and the blue one is blueberry jam.

several /ˈsevrəl/ adj. 几个的
pineapple /ˈpaɪnæpl/ n. 菠萝
durian /ˈdʊəriən/ n. 榴莲
scoop /skuːp/ v. 用勺舀
jam /dʒæm/ n. 果酱

ice cream /aɪs kriːm/ n. 冰淇淋
strawberry /ˈstrɔːbəri/ n. 草莓
experience /ɪkˈspɪəriəns/ n. 经验
a ball of /ə bɔːl əv/ 一团……

flavor /ˈfleɪvə/ n. 味道，口味
vanilla /vəˈnɪlə/ n. 香草
forget /fəˈget/ v. 忘记
watermelon /ˈwɔːtəmelən/ n. 西瓜

6. 你做的水果拼盘真漂亮啊。 The fruit platter you made is very beautiful.
7. 把水果切成片放在一个大浅盘上。 Slicing and arranging the fruits on a platter.
8. 果汁都是鲜榨的。 The fruit juices are all freshly squeezed.
9. 冰镇酸梅汁可以去腻。 The Iced sour plum drink helps to get rid of greasy taste.
10. 我们有灌装的芒果汁。 We have canned mango juice.
11. "清甜蔬菜和柠檬"是我们最好卖的果汁。 "Sweet Greens and Lemon" is our best-selling juice.

fruit platter /fruːt ˈplætə(r)/ 水果拼盘
arranging /əˈreɪndʒɪŋ/ v. 整理（arrange 的现在分词），这里指摆盘
iced /aɪst/ adj. 冰的
get rid of /get rɪd əv/ 去除
canned /kænd/ adj. 灌装的
greens /griːnz/ n. 绿叶蔬菜

slicing /ˈslaɪsɪŋ/ v. 切片（slice 的现在分词）
freshly squeezed /ˈfreʃli skwiːzd/ 鲜榨的
plum /plʌm/ n. 李子，梅子
greasy /ˈgriːzi/ adj. 油腻的
mango /ˈmæŋgəʊ/ n. 芒果
lemon /ˈlemən/ n. 柠檬

12. 果汁包括麦草、苹果、柠檬、青柠和羽衣甘蓝。 The juice includes wheat grass, apple, lemon, lime, and kale.
13. 我已经学会做巧克力脆片饼干。 I've learnt how to make chocolate chip cookies.
14. 我们来做苹果派和栗子蛋糕。 Let's make apple pie and chestnut cake.
15. 我们需要柠檬精、香草精和可可粉。 We need lemon extract, vanilla extract, and cocoa powder.
16. 今天我们有好几种水果塔：法式苹果塔、椰丝塔、杂莓塔。 Today we offer several fruit tarts, the French apple tart, coconut tart, and mixed berries tart.
17. 一定要试试草莓拿破仑酥。 The strawberry Napoleon is a must-try.

wheat grass /wiːt grɑːs/ 麦草
kale /keɪl/ n. 羽衣甘蓝
chocolate chip /ˈtʃɒklət tʃɪp/ 巧克力片
chestnut /ˈtʃesnʌt/ n. 板栗，栗子
cocoa powder /ˈkəʊkəʊ ˈpaʊdə(r)/ 可可粉
coconut /ˈkəʊkənʌt/ n. 椰子
Napoleon 此处为 Napoleon pastry /nəˈpəʊliən ˈpeɪstri/ 拿破仑酥的缩写

lime /laɪm/ n. 青柠，莱姆
learnt /lɜːnt/ v. 学习，了解到（learn 的过去分词）
cookie /ˈkʊki/ n. 饼干
extract /ˈekstrækt/ n. 精华，萃取物
tart /tɑːt/ n. 果馅饼
berries /ˈberiz/ n. 莓果，浆果（berry 复数）

Task 任务　单词背诵

背诵附录表里的水果词汇。同学们以组为单位，互相抽查词汇记忆。

 Dialogues 对话

1. At the Juice Bar 在果汁档

（初级厨师：Steve，客人：Bella)

Steve: Would you like something to drink?
Bella: I want some freshly squeezed juice.
Steve: What kind of fruit juice would you like? What flavor would you like? Like watermelon, lemon, mango, strawberry, pineapple, etc.
Bella: I'd like to have a glass of watermelon juice, without ice.
Steve: Alright, just a moment, please.

词汇

freshly squeezed juice /ˈfreʃli skwiːzd dʒuːs/ 鲜榨果汁
kind /kaɪnd/ n. 种类
flavor /ˈfleɪvə/ n. 味道，口味
watermelon /ˈwɔːtəmelən/ n. 西瓜
lemon /ˈlemən/ n. 柠檬
mango /ˈmæŋɡəʊ/ n. 芒果
strawberry /ˈstrɔːbəri/ n. 草莓
pineapple /ˈpaɪnæpl/ n. 菠萝
etc. /ˌet ˈsetərə/ abbr. 等等，以及其他 (et cetera)
a glass of /ə ɡlɑːs əv/ 一杯
without /wɪˈðaʊt/ prep. 不和……在一起
ice /aɪs/ n. 冰
moment /ˈməʊmənt/ n. 片刻，瞬间

中文

Steve：点些什么喝的吗？
Bella：我想要鲜榨果汁。
Steve：你要喝什么水果汁？你想要什么口味的？比如西瓜、柠檬、芒果、草莓、菠萝，等等。
Bella：要一杯西瓜汁，不加冰。
Steve：好的，请稍等。

注释

（1）常见的饮料词汇。

果汁 fruit juice　　　　　　　　　茶 tea
鲜榨果汁 freshly squeezed juice　　红茶 black tea
橙汁 orange juice　　　　　　　　奶茶 milk tea
汽水 soda　　　　　　　　　　　咖啡 coffee
可乐 coke　　　　　　　　　　　拿铁咖啡 latte
雪碧 sprite　　　　　　　　　　　酸奶 yogurt
柠檬水 lemonade　　　　　　　　矿泉水 mineral water

> （2）英文在表达数量时通常使用量词词组。比如：
> 一杯可乐 a glass of coke　　　　一盒牛奶 a box of milk
> 一杯茶 a cup of tea　　　　　　一瓶矿泉水 a bottle of mineral water
> 一杯咖啡 a mug of coffee　　　　一罐啤酒 a can of beer
> 其中，"glass"是玻璃杯，"cup"是带柄小杯，"mug"是带柄的大杯，也称为马克杯。

2. Mixing the Baking Materials 将烘焙材料混合

（初级厨师：Steve，饼房领班：Rosa）

Rosa: The dough is proved. Now I'll add the ingredients to the dough.
Steve: Some fruits?
Rosa: That's right. We shall use fresh raspberries to add flavors to the dough. And some mixed peel and flaked almonds.
Steve: Is that all?
Rosa: And add some lemon and orange zest. At last, add some juice of lemon in there as well.
Steve: Lovely. Let me do the kneading. I'm enjoying that.
Rosa: We don't have to stick to this choice of nuts and fruits, let's try raisins and cinnamon next time.

词汇

dough /dəʊ/ n. 生面团
proved /pruːvd/ adj. 发酵了的
add /æd/ v. 加，增加
ingredient /ɪnˈɡriːdiənts/ n. 材料
raspberries /ˈrɑːzbəriz/ n. 树莓，覆盆子（raspberry 复数形式）
mixed /mɪkst/ adj. 混合的
peel /piːl/ n. 果皮
flaked /fleɪkt/ adj. 薄片的
almond /ˈɑːmənd/ n. 杏仁
zest /zest/ n. 柑橘外皮
lemon zest /ˈlemən zest/ 柠檬皮碎末
orange zest /ˈɔrɪndʒ zest/ 橘子皮碎末
lovely /ˈlʌvli/ adj. 漂亮的，可爱的，令人愉快的
kneading /ˈniːdɪŋ/ n. 捏合、揉捏（knead 的现在分词）
stick to /stɪk tu/ 坚持
nuts /nʌts/ n. 坚果
raisin /ˈreɪzn/ n. 葡萄干
cinnamon /ˈsɪnəmən/ n. 桂皮

中文

Rosa：面团发酵好了。现在我要将配料添加到面团中。
Steve：水果吗？
Rosa：是的。我们要用新鲜的树莓为面团添加味道。还有一些混合的果皮和杏仁片。
Steve：就这么多吗？
Rosa：还有柠檬皮和橘子皮碎末。最后，加点柠檬汁进去。
Steve：太棒了。我来揉吧。我喜欢这个。
Rosa：我们不必拘泥于坚果和水果，下次试试葡萄干和肉桂。

注释

（1）描述制作面食时常用的动词。

mix 和面	knead 揉面	prove 发酵
roll 擀面	flatten 擀平	crosshatch 画交叉线
shape 塑型	prove again 再发酵	bake 烤

（2）面粉混合了酵母、盐、油等以后成为生面团，它的英文是"dough"；生面团发酵成型以后则称为"pastry"。以下是常见句型：

用双手将面粉和成面团 work flour into dough with the hands

揉面 knead the dough

擀面 roll the dough

擀平面团 flatten the dough

面皮成型 shape the pastry

面皮上画交叉线 crosshatch the pastry

* 记住这些制作面食时常用的动词，在需要用英文来表达面点制作的情景时，就难不倒你了。

3. Breakfast Breads and Pastries 早餐的面包与酥饼

（初级厨师：Steve，饼房领班：Rosa）

Rosa: It is our job to prepare bakery products. How many types of breakfast breads and pastries do you know?

Steve: I can name all: toast, bun, muffin, croissant, Danish pastry, baguette, brownie, doughnut, pancake, waffle, bagel, soft roll, hard roll...

词汇

bakery /ˈbeɪkəri/ *n.* 面包店，饼房
product /ˈprɒdʌkt/ *n.* 产品
type /taɪp/ *n.* 种类，类型
breads /bredz/ *n.* 各种面包（bread 的复数形式）
pastries /ˈpeɪstriz/ *n.* 点心；甜点；酥皮糕点（pastry 的复数）
name /neɪm/ *v.* 叫出；称呼
toast /təʊst/ *n.* 吐司
bun /bʌn/ *n.* 小圆面包
muffin /ˈmʌfɪn/ *n.* 松饼，玛芬
croissant /krəˈsɑːnt/ *n.* 牛角包
Danish pastry /ˈdeɪnɪʃ ˈpeɪstri/ 丹麦酥
baguette /bæˈɡet/ *n.* 法式长棍
brownie /ˈbraʊni/ *n.* 核仁巧克力饼，布朗尼
doughnut /ˈdəʊnʌt/ *n.* 甜甜圈
bagel /ˈbeɪɡl/ *n.* 贝果面包
soft /sɒft/ *adj.* 软的
roll /rəʊl/ *n.* 小面包；卷

Rosa: You got this. Our guests care about the quality of the breakfast and the variety of items served up.

Steve: Yes. We all love fluffy pancakes and flaky croissants, the fresh-baked or toasted breads.

Rosa: That's true. Come on, let's work on this.

quality /ˈkwɒləti/ **n**. 质量	
variety /vəˈraɪəti/ **n**. 多样；变化	
item /ˈaɪtəm/ **n**. 物品，商品	
fluffy /ˈflʌfi/ **adj**. 蓬松的；松软的	
flaky /ˈfleɪki/ **adj**. 薄而易剥落的，起酥的	
fresh-baked /freʃ beɪkt/ 新鲜烤出炉的	
toasted /ˈtəʊstɪd/ **adj**. 烤的	

中文

Rosa: 准备焙烤食品是我们的工作。你知道多少种早餐面包和酥饼？

Steve: 我可以全都说出来：吐司、面包、松饼、牛角包、丹麦酥、法式长棍、布朗尼、甜甜圈、煎饼、华夫饼、贝果、软面包、硬面包……

Rosa: 可以啊。我们的客人在乎早餐的质量和提供的食物种类多少。

Steve: 是的。我们都喜欢蓬松的煎饼和起酥的牛角包，以及新鲜出炉的面包。

Rosa: 没错。来吧，让我们继续努力。

注释

从街头小巷的杂货店到星级酒店的精致下午茶，西式点心已成为日常食品。其中有些食品因为商家使用了不同的译名而引起大众的困惑。例如：

英文词汇	通用译名	其他译名
cheese	奶酪	芝士、干酪、起司
cream	奶油	稀奶油，忌廉（粤语音译）
toast	吐司	土司、烤面包、多士（粤语音译）
croissant	牛角包	可颂、羊角包
doughnut	甜甜圈	炸面圈、冬甩（粤语音译）
pancake	煎饼	烤薄饼、班戟（粤语音译）
muffin	松饼	玛芬、杯状小松糕
bagel	贝果	百吉饼、硬面包圈
souffle	蛋奶酥	舒芙里、舒芙蕾
yogurt	酸奶	优格乳

直接学习外来食品的英文词汇能帮助同学们在学习烘焙知识之余更高效地识别菜单和食谱里的菜名及材料。

积累了相当的水果英文词汇可以更自如地应对烘焙工作。以下列举了常见水果及坚果的中英对照词汇，必背。

Fruits and Nuts			
英文	中文	英文	中文
apple /ˈæp(ə)l/	苹果	pear /peə(r)/	梨
banana /bəˈnɑːnə/	香蕉	orange /ˈɒrɪndʒ/	橙
lemon /ˈlemən/	柠檬	lime /laɪm/	青柠，莱姆
tangerine /ˌtændʒəˈriːn/	柑橘（皮硬的，微酸）	mandarin /ˈmændərɪn/	柑橘（皮软易剥的，更甜）
kumquat /ˈkʌmkwɒt/	金橘	kiwi fruit /ˈkiːwi fruːt/	奇异果
pomelo /ˈpɒmɪləʊ/	柚子	grapefruit /ˈgreɪpfruːt/	葡萄柚，西柚
starfruit /ˈstɑː(r)ˌfruːt/	杨桃	jackfruit /ˈdʒækfruːt/	菠萝蜜，木菠萝
Passion fruit /ˈpæʃən fruːt/	百香果	dragon fruit /ˈdrægən fruːt/ pitaya /pɪˈtaɪə/	火龙果
strawberry /ˈstrɔːbəri/	草莓	blueberry /ˈbluːbəri/	蓝莓
blackberry /ˈblækbəri/	黑莓	raspberry /ˈrɑːzbəri/	覆盆子，树莓
cranberry /ˈkrænbəri/	蔓越莓，小红莓	waxberry /ˈwæksbəri/	杨梅
cherry /ˈtʃeri/	樱桃	olive /ˈɒlɪv/	橄榄
watermelon /ˈwɔːtəmelən/	西瓜	muskmelon /ˈmʌskˌmelən/	香瓜
sweet melon /swiːt ˈmelən/ honeydew melon /ˈhʌnidjuː ˈmelən/	哈蜜瓜	papaya /pəˈpaɪə/	木瓜
durian /ˈdʊəriən/	榴莲	pineapple /ˈpaɪnæpl/	菠萝
custard apple /ˈkʌstəd ˈæpl/	释迦果	wax apple /wæks ˈæpl/	莲雾
grape /greɪp/	葡萄	raisin /ˈreɪzn/	葡萄干
apricot /ˈeɪprɪˌkɒt/	杏子	almond /ˈɑːmənd/	杏仁

（续）

Fruits and Nuts			
英文	中文	英文	中文
guava /ˈgwɑːvə/	番石榴，芭乐	pomegranate /ˈpɒmɪgrænɪt/	石榴
persimmon /pəˈsɪmən/	柿子	sugar cane /ˈʃʊgə(r) keɪn/	甘蔗
fig /fɪg/	无花果	haw /hɔː/	山楂
jujube /ˈdʒuːdʒuːb/	枣子	date /deɪt/	海枣，椰枣
loquat /ˈləʊkwɒt/	枇杷，卢橘	longan /ˈlɒŋ(ə)n/	龙眼
wampee /wɒmˈpiː/	黄皮	lychee\litchi /ˈlaɪtʃi/	荔枝
coconut /ˈkəʊkənʌt/	椰子	peanut /ˈpiːnʌt/	花生
walnut /ˈwɔːlnʌt/	核桃	chestnut /ˈtʃesnʌt/	栗子
cashew nut /ˈkæʃuːnʌt/	腰果	betelnut /ˈbiːtəlnʌt/	槟榔
mango /ˈmæŋgəʊ/	芒果	mangosteen /ˈmæŋgəstiːn/	山竹
peach /piːtʃ/	桃子	nectarine /ˈnektəriːn/	油桃
plum /plʌm/	李子，梅子，布林	avocado /ˌævəˈkɑːdəʊ/	牛油果，鳄梨

练习：掌握常见水果词汇

完成以下食物翻译练习，并背诵词汇。

序号	中文	英文	序号	中文	英文	序号	中文	英文
1	香蕉		3	番石榴		5	西瓜	
2	菠萝		4	草莓		6	柚子	

（续）

序号	中文	英文	序号	中文	英文	序号	中文	英文
7	柠檬		17	哈密瓜		27	牛油果	
8	椰子		18	葡萄柚		28	树莓	
9	桃子		19	青柠		29	蔓越莓	
10	荔枝		20	核桃		30	柑橘	
11	榴莲		21	油桃		31	奇异果	
12	香草冰淇淋		22	龙眼		32	杨梅	
13	苹果		23	木瓜		33	杨桃	
14	火龙果		24	杏仁蛋糕		34	李子	
15	石榴		25	芒果		35	菠萝蜜	
16	樱桃		26	百香果		36	草莓布丁	

职业提示

人民至上，安全意识

近年来，政府出台了许多政策来守护"舌尖上的安全"。食品安全事故的预防包括厨房卫生、厨具消毒、添加剂的使用剂量等。我们餐饮从业人员更应该把顾客的饮食安全放在首位，坚守职业道德，遵守厨房安全管理规章制度。积极推进健康中国建设，不断提升餐饮服务的品质和水平，为保障人民健康、促进经济发展、建设美好中国做出积极的贡献。

Part 3
Dinner
正 餐

Unit 8   Quick Western Meals 西餐简餐／078

Unit 9   Menu of Western Food 西餐菜单／093

Unit 10   Menu of Chinese Food 中餐菜单／106

Unit 8 Quick Western Meals
西餐简餐

西餐里简便的一顿饭菜通常指沙拉、汉堡包、比萨、意大利面、三明治等。普通的西式餐厅、咖啡店、酒吧以及星级酒店都会供应西餐简餐。本单元列举了一些常见西式简餐的英文词汇及其句型,希望同学们通过学习能将同类型的表达进行举一反三。

Learning Objectives 学习目标

❈ Know common quick meal vocabularies.
 了解常见的便餐词汇。
❈ Learn some expressions of making sandwich, salad and pasta.
 学习一些制作三明治、沙拉和意大利面的表达。

扫码看视频

Basic Terms 基础词汇

Unit 8　Quick Western Meals　西餐简餐 /079

卷心菜沙拉	恺撒沙拉	蜜瓜火腿
/ˈkəʊlslɔː ˈsæləd/	/ˌsiːzə ˈsæləd/	
coleslaw salad	**caesar salad**	**ham and melon**

沙拉酱	蛋黄酱	番茄酱
/ˈsæləd dresɪŋ/	/ˌmeɪəˈneɪz/	/ˈketʃəp/
salad dressing	**mayonnaise**	**ketchup**

海鲜至尊比萨	海陆双霸王比萨	千层面 / 烤宽面
		/ləˈzænjə/
seafood supreme pizza	**surf & turf delicacy pizza**	**lasagne**

意式细面	通心粉	尖管通心粉
/spəˈgeti/	/mækəˈrəʊni/	/ˈpeneɪ/
spaghetti	**macaroni**	**penne**

意大利螺旋面 意式细面 蝴蝶结面
细、短

/fʊˈzili/ **fusilli**　　/ˌvɜːmɪˈtʃeli/ **vermicelli**　　/fɑːrˈfæli/ **farfalle**

意大利宽面条 意式扁平面

/ˌfetʊˈtʃiːni/ **fettuccine**　　/lɪŋˈɡwiːni/ **linguine**

Functional Expressions 实用表达

请流利地朗读以下句子，并做到中英互译。

1. 水果沙拉非常新鲜多汁。　　The fruit salad is very fresh and juicy.
2. 要确保水果都成熟了。　　Make sure the fruits are ripe.
3. 您的沙拉要什么酱汁？　　What kind of dressing would you like on your salad?
4. 在沙拉上淋上酱汁。　　Top salad with dressing.
5. 把酱汁放在菜的边上。　　Serve the dressing on the side.

fresh /freʃ/ *adj.* 新鲜的　　juicy /ˈdʒuːsi/ *adj.* 多汁的　　ripe /raɪp/ *adj.* 熟的，成熟的
dressing /ˈdresɪŋ/ *n.* （拌制色拉用的）调料　　top /tɒp/ *v.* 放在……的上面
serve /sɜːv/ *v.* 提供；端上　　side /saɪd/ *n.* 侧面，旁边

6. 最好是在上面放点千岛酱。　　It's best topped with Thousand Island dressing.
7. 三明治里有洋葱、辣椒和奶酪。　　The sandwich is filled with onion, peppers, and cheese.
8. 我认为最好的汉堡包肉饼是肥的牛肉饼。　　I think the best hamburger patty is the fatty ground beef.
9. 汉堡包和薯条是绝配。　　Hamburger pairs perfectly with French fries.
10. 做一个新鲜的番茄酱浇在意面上。　　Make a fresh tomato sauce to top our pasta.

topped with /tɒpt wɪð/ 淋上……（topped 是 top 的过去分词）
Thousand Island dressing /ˈθaʊznd ˈaɪlənd ˈdresɪŋ/ 千岛酱（由蛋黄酱与番茄酱混合制成）

sandwich /ˈsænwɪdʒ/ n. 三明治	fill with /fɪl wɪð/ 填满，填充	onion /ˈʌnjən/ n. 洋葱
pepper /ˈpepə(r)/ n. 辣椒	cheese /tʃiːz/ n. 奶酪，芝士	hamburger /ˈhæmbɜːɡə(r)/ n. 汉堡包
patty /ˈpæti/ n. 肉饼；小馅饼	hamburger patty /ˈhæmbɜːɡə(r) ˈpæti/ 汉堡包肉饼	
fatty /ˈfæti/ adj. 富含脂肪的	ground beef /ˌɡraʊnd ˈbiːf/ 碎牛肉，牛肉糜，牛肉酱	
pair with /peə(r) wɪð/ 与……成对，与……配对		perfectly /ˈpɜːfɪktli/ adv. 完美地
french fries /frentʃ fraɪz/ 炸薯条	sauce /sɔːs/ n. 酱汁；沙司	pasta /ˈpæstə/ n. 意大利面

11. 意大利面是我们许多菜的通用底料。 Pasta is a versatile base for many of our dishes.
12. 你想尝尝我特制的意大利肉酱面吗？ Would you like to try my exclusive spaghetti bolognese?
13. 你想如何准备你的沙拉由你决定。 How you want to prepare your salad is up to you.
14. 你可以用你手头上的任何肉类和蔬菜来做。 You can add whatever meat and vegetable you have on hand.
15. 有几种口味。 There are several flavors.

versatile /ˈvɜːsətaɪl/ adj. 通用的，万能的	base /beɪs/ n. 基底；基础
exclusive /ɪkˈskluːsɪv/ adj. 独有的	spaghetti bolognese /spəˈɡeti bɒləˈneɪz/ n. 意大利肉酱面
whatever /wɒtˈevə(r)/ det. 任何事物	several /ˈsevrəl/ adj. 几个的

16. 恺撒沙拉以生菜、面包丁和奶油蒜调味料为基础。 Caesar salad has lettuce, croutons, and creamy garlic dressing as the foundation.
17. 配料包括一片生菜和一个白煮蛋。 The ingredients include a bed of lettuce and a hard-boiled egg.
18. 我们只需要使用新鲜的蔬菜及好品质的肉类。 We only need fresh vegetables and good quality meats.
19. 把洋葱圈蘸面糊后再油炸。 Dip the onion rings into the batter and deep fry.
20. 这个意大利面煮得很有嚼劲。 The pasta is cooked al dente.

caesar salad /ˌsiːzə ˈsæləd/ 恺撒沙拉	lettuce /ˈletɪs/ n. 生菜 crouton /ˈkruːtɒn/ n. 油炸面包丁
creamy /ˈkriːmi/ adj. 含乳脂的	garlic dressing /ˈɡɑːlɪk dresɪŋ/ 大蒜酱
foundation /faʊnˈdeɪʃn/ n. 基础	ingredient /ɪnˈɡriːdiənt/ n. 原料，成分
a bed of /ə bed əv/ 一片……	hard-boiled egg /hɑːd bɔɪld eɡ/ 煮得较熟的水煮蛋
quality /ˈkwɒləti/ n. 质量	dip /dɪp/ v. 浸，蘸 onion rings /ˈʌnjən rɪŋz/ 洋葱圈
batter /ˈbætə(r)/ n. 面糊	deep fry /ˌdiːp ˈfraɪ/ 油炸
al dente /ælˈdenteɪ/ adj.（食物，尤指面食）筋道的，有韧性耐咀嚼的	

Task 任务　小组活动

讨论：1. 汉堡包和三明治有什么区别？
　　　2. 意大利面有哪些常见的品种？

翻译：1. 鸡肉奶酪汉堡包与鸡肉奶酪三明治
2. 肉酱意面与奶油意面

Dialogues 对话

1. Making a Sandwich 做一份三明治

（初级厨师：Steve，行政副厨师长：Jackson）

Jackson:	A sandwich is a quick meal. What's your favorite condiments that you want on your sandwich?
Steve:	Cheese and meat. What condiments to put in between 2 slices of bread?
Jackson:	Actually we can experiment with different ingredients. Grab some mustard, stack up the meat and cheese.
Steve:	OK. On top of the mustard goes the meat. Then Cheese.
Jackson:	That's right. At your stage, you can follow a recipe to make a well-known classic sandwich.
Steve:	Yes, chef.

词汇

sandwich /ˈsænwɪdʒ/ n. 三明治
quick meal /kwɪk miːl/ 快餐，便饭
favorite /ˈfeɪvərɪt/ adj. 最喜欢的
condiment /ˈkɒndɪmənt/ n. 调味品
2 slices of 两片……
slice /slaɪs/ n. 片，薄片
actually /ˈæktʃuəli/ adv. 实际上；事实上
experiment /ɪkˈsperɪmənt/ v. 尝试，试验
grab /græb/ v. 拿；攫取
mustard /ˈmʌstəd/ n. 芥末
stack up /stækʌp/ 叠放；堆积
stage /steɪdʒ/ n. 阶段，时期；状态
follow /ˈfɒləʊ/ v. 遵循；跟随
recipe /ˈresəpi/ n. 食谱
well-known /ˌwelˈnəʊn/ adj. 著名的
classic /ˈklæsɪk/ adj. 经典的

中文

Jackson:	三明治属于便餐。三明治你最喜欢放什么佐料？
Steve:	奶酪和肉。在两片面包之间放入什么调味料好？
Jackson:	其实我们可以尝试不同的配方。拿些芥末，把肉和奶酪堆起来。
Steve:	好的。芥末上面放肉。然后放奶酪。
Jackson:	对。在你这个阶段，你可以按照食谱制作一份著名的经典三明治。
Steve:	好的，厨师长。

注释

（1）"三明治"是英文"sandwich"的音译。它是一种典型的西方食品，用两片面包夹几片肉和芝士、各种调料制作而成，吃法简便。传说名字源于18世纪英国一位爱打桥牌、名为 Sandwich 的伯爵。Sandwich 是所有"面包夹东西"的总称。比如：

俱乐部三明治　club sandwich
金枪鱼三明治　tuna sandwich
烤牛肉三明治　roast beef sandwich

全麦三明治 whole wheat sandwich

鸡肉恺撒沙拉卷 chicken caesar salad wrap

（2）三明治家族里最知名的就是"俱乐部三明治"（club sandwich）了，它也称为公司三明治。这是一款在北美非常传统的三明治，很多快餐店和小餐馆都能买到。它由煎蛋（fried eggs）、火腿（ham）、蔬菜（vegetables）、奶酪（cheese）、烟肉（bacon）和番茄（tomato）等各式食材制作而成。摆盘通常为切成四等份以后用牙签穿好。

2. Making a Chef's Salad 做一份厨师沙拉

（初级厨师：Steve，行政副厨师长：Jackson）

Jackson: When making a chef's salad, the foundation is a bed of lettuce topped with meats, cheese, tomatoes, and hard-boiled eggs.
Steve: What else can we put into?
Jackson: Then just use what you have on hand. It looks different on menus of different restaurants. I always top it with a grilled chicken breast.
Steve: I'd like to top it with diced ham.
Jackson: It could work. At last, serve the dressing on the side.

词汇

chef's salad /ʃefs sæləd/ 厨师特制沙拉，厨师沙拉
foundation /faʊnˈdeɪʃn/ n. 基础
a bed of /ə bed əv/ 一片……
lettuce /ˈletɪs/ n. 生菜
hard-boiled egg /hɑːd bɔɪld eg/ 煮得较熟的水煮蛋
else /els/ adv. 其他的；别的
on hand /ɒn hænd/ 现有的，手头上的
menu /ˈmenjuː/ n. 菜单
restaurant /ˈrestrɒnt/ n. 餐厅
grilled /grɪld/ adj. 烤的
chicken breast /ˈtʃɪkɪn brest/ 鸡胸肉
diced /daɪst/ adj. 切丁的
ham /hæm/ n. 火腿
serve /sɜːv/ v. 提供；端上
side /saɪd/ n. 侧面，旁边

中文

Jackson: 制作厨师沙拉时，沙拉底是生菜，上面放有肉、奶酪、西红柿和全熟的白煮蛋。
Steve: 我们还能放什么吗？
Jackson: 然后就使用你手头上现有的食材。这道菜在不同餐厅的菜单上看起来都不一样。我总是在上面加一块烤鸡胸肉。
Steve: 我要加火腿丁。
Jackson: 可以。最后，将酱汁放在旁边。

注释

沙拉是常见的西餐菜品，可作为头盘或主菜。在为沙拉选取酱汁时可以使用如下句型：

what kind of dressing would you like on your salad? 您的沙拉要什么酱汁？

top the salad with Thousand Island dressing　　在沙拉上面放点千岛酱
pour vinaigrette over the greens　　把醋汁倒在蔬菜上
drizzle some honey　　淋上一点蜂蜜
toss to coat　　搅拌均匀
sprinkle cheese over the top　　在上面撒上奶酪

制作沙拉时讲究原材料的新鲜及酱汁的搭配。以下是两种常见的沙拉及搭配的酱汁：

Greek salad 希腊沙拉
　　配料：cucumbers 黄瓜
　　　　　onion 洋葱
　　　　　olives 橄榄
　　　　　peppers 甜椒
　　　　　tomatoes 西红柿
　　　　　feta cheese 菲达奶酪／羊奶奶酪

vinaigrette 油醋汁
　　配料：olive oil 橄榄油
　　　　　balsamic vinegar 香醋
　　　　　Dijon mustard 第戎芥末
　　　　　honey 蜂蜜
　　　　　black pepper 黑胡椒
　　　　　garlic 大蒜

caesar salad 恺撒沙拉
　　配料：lettuce 生菜
　　　　　croutons 炸面包粒
　　　　　parmesan Cheese 帕尔马奶酪

caesar dressing 恺撒酱
　　配料：anchovy 凤尾鱼
　　　　　mayonnaise 蛋黄酱
　　　　　Parmesan Cheese 帕尔马奶酪
　　　　　garlic 大蒜

3. Making a Spaghetti Bolognese 做一份意大利肉酱面

（初级厨师：Steve，行政副厨师长：Jackson）

Jackson: Spaghetti bolognese is an Italian classic. Now let's start making the bolognese sauce, the ragu.
Steve: I know ragu is a sauce made from tomatoes and ground beef or meat.
Jackson: Brilliant! To be brief, fry the onions, carrots, and celery for 10 minutes. Add ground beef, milk, and white wine, stew them for 3 hours.
Steve: That's quite a long time.
Jackson: That makes the sauce rich, creamy, and silky smooth.
Steve: And for the spaghetti?

词汇

spaghetti bolognese /spəˈgeti bɒləˈneɪz/ 意大利肉酱面
Italian /ɪˈtæliən/ *adj.* 意大利的
classic /ˈklæsɪk/ *n.* 优秀的典范；杰作
sauce /sɔːs/ *n.* 酱汁
ragu /ræˈguː/ *n.* 肉酱（意式烹调）
made from /meɪd frəm/ 由……做成
ground beef /ˌɡraʊnd ˈbiːf/ 搅碎的牛肉
brilliant /ˈbrɪliənt/ *adj.* 极棒的；优秀的
brief /briːf/ *adj.* 简短的　*n.* 概要
fry /fraɪ/ *v.* 煎，炒
carrot /ˈkærət/ *n.* 胡萝卜
celery /ˈseləri/ *n.* 芹菜
white wine /ˌwaɪt ˈwaɪn/ 白葡萄酒
stew /stjuː/ *v.* 炖，焖

Jackson:	On the other side, boil dry spaghetti for 10 minutes. Don't forget to add olive oil and salt.	rich /rɪtʃ/ *adj.* 浓郁的，含有很多脂肪的 creamy /ˈkriːmi/ *adj.* 含乳脂的 silky /ˈsɪlki/ *adj.* 柔滑的 smooth /smuːð/ *adj.* 光滑的 boil /bɔɪl/ *v.* 煮沸 olive oil /ˈɒlɪv ɔɪl/ 橄榄油 salt /sɔːlt/ *n.* 盐 al dente /ælˈdenteɪ/ *adj.* 有嚼劲的 turn off the heat /tɜːn ɒf ðə hiːt/ 关火 plate /pleɪt/ *v.* 装碟
Steve:	The spaghetti is cooked al dente.	
Jackson:	Turn off the heat for both the spaghetti and ragu sauce. Plate up!	

中文

Jackson: 意大利肉酱面是意大利经典菜式。现在来制作肉酱，也叫作拉古肉酱。
Steve: 我知道拉古肉酱是用西红柿、牛肉沫或其他肉沫制成的酱。
Jackson: 很棒！简单来说，将洋葱、胡萝卜和芹菜炒 10 分钟。加入牛肉沫、牛奶和白葡萄酒，炖 3 小时。
Steve: 时间挺长的。
Jackson: 这样酱汁才会丰厚醇香，奶味浓郁，口感柔滑。
Steve: 那意大利面呢？
Jackson: 另一边，将意大利面煮 10 分钟。不要忘记放橄榄油和盐。
Steve: 意面这样煮有嚼劲。
Jackson: 同时关掉意面和肉酱的火。装碟！

注释

（1）pasta 是意大利面的统称。意大利语，泛指所有源自意大利的面食。真要算起来，意大利面的种类可达 500 多种。以下列举几种常见的意式面条：

长型细面条	spaghetti	/spəˈgeti/
宽条面	fettuccine	/ˌfetuˈtʃiːni/
扁平面	linguine	/lɪŋˈgwiːni/
通心粉	maccheroni	/ˌmækəˈrəuni/
长通心粉	penne	/ˈpeneɪ/
蝴蝶结面	farfalle	/faːrˈfæli/
螺旋粉	fusilli	/fʊˈziːli/
千层面	lasagne	/ləˈzanjə/
细面条	vermicelli	/ˌvɜːmiˈtʃeli/
方饺	ravioli	/ˌræviˈəuli/

（2）pizza 也源自意大利，是一种涂有奶酪和番茄酱的意大利式有馅烘饼。比如：意大利香肠辣椒比萨（Italian sausage and peppers pizza）、海鲜至尊比萨（seafood supreme pizza），及乳酪大会（cheese lovers pizza）。

（3）蔬菜词汇是必须要掌握的。以下列举了常见蔬菜的中英对照词汇，必背。

Vegetables			
greens /griːnz/	绿叶蔬菜	escarole /ˈeskərəʊl/	阔叶菊苣，茅菜，沙拉菜
spinach /ˈspɪnɪtʃ/	菠菜	water spinach /ˈwɔːtə(r) ˈspɪnɪtʃ/	空心菜
lettuce /ˈletɪs/	生菜，莴苣	Chinese lettuce /ˌtʃaɪˈniːz ˈletɪs/	莜麦菜
cabbage /ˈkæbɪdʒ/	卷心菜	Chinese cabbage /ˌtʃaɪˈnlːz ˈkæbɪdʒ/	大白菜
leek /liːk/	韭葱	Chinese flowering cabbage /ˌtʃaɪˈniːz ˈflaʊərɪŋ ˈkæbɪdʒ/	菜心
Chinese Chives /ˌtʃaɪˈniːz tʃaɪvz/	韭菜	Chinese Chives blossom /ˌtʃaɪˈniːz tʃaɪvz ˈblɒsəm/	韭菜花
chive /tʃaɪv/	葱叶；韭菜	shallot /ʃəˈlɒt/	红葱头
cilantro /siˈlæntrəʊ/ coriander /kɒriˈændə(r)/	香菜，芫荽	scallion /ˈskæliən/ spring onion /sprɪŋ ˈʌnjən/	葱
garlic /ˈɡɑːlɪk/	大蒜，蒜头	garlic sprout /ˈɡɑːlɪk spraʊt/	蒜苗
onion /ˈʌnjən/	洋葱	ginger /ˈdʒɪndʒə/	姜
pepper /ˈpepə(r)/	胡椒，辣椒	hot pepper /hɒt ˈpepə(r)/ chilli /ˈtʃɪli/	辣椒（辣）
broccoli /ˈbrɒkəli/	西蓝花	cauliflower /ˈkɒliflaʊə(r)/	菜花，花椰菜
corn /kɔːn/	玉米	tomato /təˈmɑːtəʊ/	西红柿，番茄
potato /pəˈteɪtəʊ/	马铃薯，土豆	sweet potato /swiːt pəˈteɪtəʊ/	红薯
carrot /ˈkærət/	胡萝卜	white radish /waɪt ˈrædɪʃ/	白萝卜
turnip /ˈtɜːnɪp/	芜菁；白萝卜（中餐）	celery /ˈseləri/	芹菜
cress /kres/	水芹	watercress /ˈwɔːtəkres/	西洋菜
taro /ˈtɑːrəʊ/	芋头	yam /jæm/	山药
cassava /kəˈsɑːvə/	木薯	lotus root /ˈləʊtəs ruːt/	莲藕
bamboo shoot /ˌbæmˈbuː ʃuːt/	竹笋	asparagus /əˈspærəɡəs/	芦笋
bean /biːn/	豆，菜豆	pea /piː/	豌豆
bean sprout /biːn spraʊt/	豆芽	soybean /ˈsɒibiːn/	黄豆
green bean /ɡriːn biːn/	豆角	mung bean /mʌŋ biːn/	绿豆
cucumber /ˈkjuːkʌmbə(r)/	黄瓜	pumpkin /ˈpʌmpkɪn/	南瓜

（续）

Vegetables			
zucchini /zuˈkiːni/	西葫芦，小青瓜	melon /ˈmelən/	瓜，甜瓜
bitter melon /ˈbɪtə(r) ˈmelən/	苦瓜	gourd /ɡʊəd/	葫芦
wax gourd /wæks ɡʊəd/ White gourd /waɪt ɡʊəd/	冬瓜	sporge gourd /spʌndʒ ɡʊəd/	丝瓜
fungus /ˈfʌŋɡəs/	菌类，菇类	edible fungus /ˈedəb(ə)l ˈfʌŋɡəs/	食用菌
black fungus /blæk ˈfʌŋɡəs/	木耳	white fungus /waɪt ˈfʌŋɡəs/	银耳
mushroom /ˈmʌʃruːm/	蘑菇	black mushroom /blæk ˈmʌʃruːm/	冬菇，香菇
needle mushroom /ˈniːdl ˈmʌʃruːm/	金针菇	aweto /ɑːˌfɑtɔː/	冬虫夏草
okra /ˈəʊkrə/	秋葵	goji berry /ˈɡəʊdʒi ˈberi/	枸杞
seaweed /ˈsiːwiːd/	海草，海带，紫菜	kelp /kelp/	海带
pickle /ˈpɪkl/	泡菜，腌制食品	pickled bean /ˈpɪkld biːn/	酸豆角
Chinese sauerkraut /ˌtʃaɪˈniːz ˈsaʊəkraʊt/	酸菜	skin /skɪn/, peel /piːl/ turnip peel /tɜːnɪp piːl/	皮 萝卜皮

胡萝卜

/ˈkærət/
carrot

白萝卜

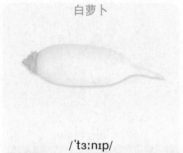

/ˈtɜːnɪp/
turnip

西红柿

/təˈmɑːtəʊ/
tomato

芦笋

/əˈspærəɡəs/
asparagus

芹菜

/ˈseləri/
celery

西蓝花

/ˈbrɒkəli/
broccoli

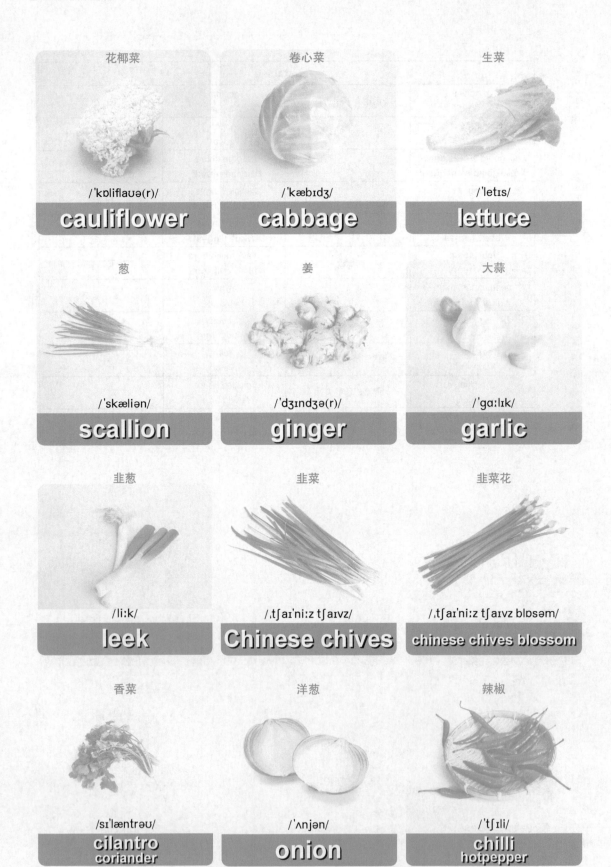

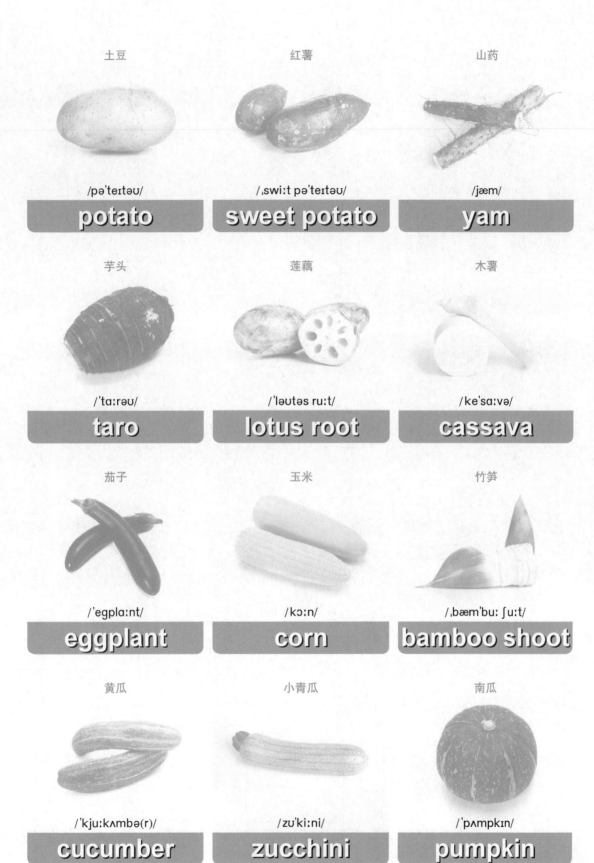

苦瓜	冬瓜	丝瓜

/ˈbɪtə(r) ˈmelən/	/wæks gʊəd/	/spʌndʒ gʊəd/
bitter melon	**wax gourd**	**sponge gourd**

西洋菜	菠菜	空心菜

/ˈwɔːtəkres/	/ˈspɪnɪtʃ/	/ˈwɔːtə(r) ˈspɪnɪtʃ/
watercress	**spinach**	**water spinach** ong choy（粤语音译）

蘑菇	金针菇	菌类

/ˈmʌʃrʊm/	/ˈniːdl ˈmʌʃrʊm/	/ˈfʌŋɡəs/
mushroom	**needle mushroom**	**fungus**

木耳	银耳	豌豆

/blæk ˈfʌŋɡəs/	/waɪt ˈfʌŋɡəs/	/piː/
black fungus	**white fungus** tremella	**pea**

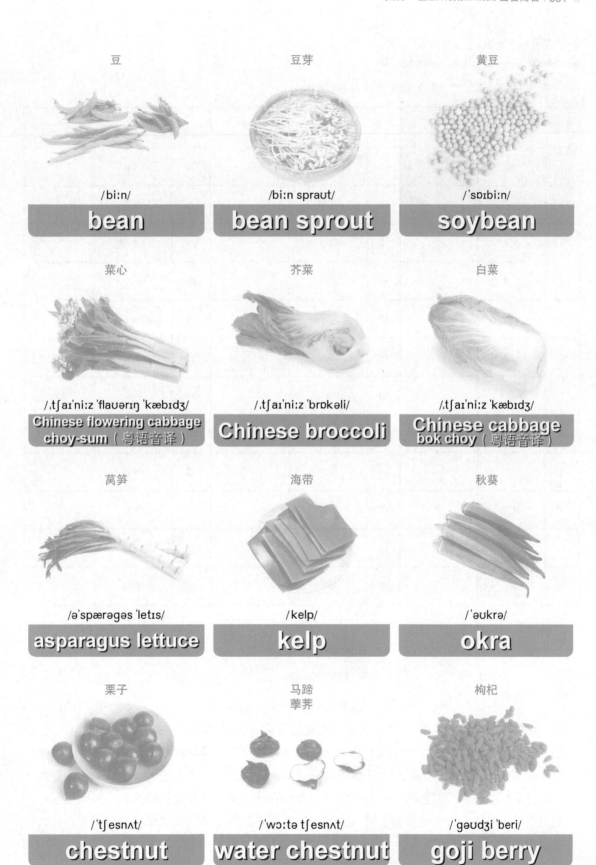

练习：掌握常见蔬菜词汇

将表内的蔬菜词汇翻译为英文，并背诵词汇。

序号	中文	英文	序号	中文	英文	序号	中文	英文
1	生菜		13	西蓝花		25	洋葱	
2	卷心菜		14	菠菜		26	芹菜	
3	玉米		15	辣椒		27	西红柿	
4	土豆		16	芦笋		28	黄瓜	
5	南瓜		17	小青瓜（西葫芦）		29	泡菜（腌制食品）	
6	豌豆		18	豆角		30	花生	
7	胡萝卜		19	白萝卜		31	芋头	
8	韭葱		20	葱		32	大蒜	
9	姜		21	韭菜		33	香菜	
10	欧芹		22	蘑菇		34	菌类	
11	豆芽		23	冬瓜		35	空心菜	
12	菜心		24	竹笋		36	莲藕	

职业提示

多彩世界，丰富文明

在党的二十大报告中，强调了推进文化强国建设的战略任务，提出了"加强中华优秀传统文化传承和创新，大力发展文化产业和文化事业"的目标和要求。只有不断吸收借鉴其他文化的优秀元素，我们才能更好地振兴中华饮食文化，为推进中华民族的伟大复兴贡献力量。西餐制作工艺简便、注重食材选用和营养价值；分餐制和公筷制有其合理性，值得借鉴。中华儿女素来就有着博大的胸怀，我们要振兴中华饮食文化，就要博采众长，为我所用。

Unit 9 Menu of Western Food
西餐菜单

　　头盘、主菜、甜点是一顿西餐正餐的基本餐序。头盘也称为开胃菜、前菜或餐前小食，以冷菜为主，分量少，精致开胃；主菜通常为肉类热菜，具有代表性的菜式有煎牛排、烤羊排及烤鸭胸等；甜点是一顿正餐的结尾，种类包含小蛋糕、水果、乳酪拼盘、冰淇淋等食物。提高西餐英文菜单的阅读能力，从掌握足够的词汇量开始。

Learning Objectives 学习目标

❄ Memorize basic terms of food ordering.
　记住点餐的基本术语。
❄ Know some classic Western dishes.
　了解一些经典的西餐菜品。

扫码看视频

Basic Terms 基础词汇

一顿三道菜的正餐

a 3-coures meal

开胃品

/ˈæpɪtaɪzə(r)/
appetizer

头盘

/ˈstɑːtə(r)/
starter

主菜

/ˌmeɪn ˈkɔːs/
main course

甜点

/dɪˈzɜːt/
dessert

牛排

/steɪk/
steak

烟熏鲑鱼

/sməukt 'sæmən/
smoked salmon

鱼子酱

/'kævia:(r)/
caviar

鹅肝酱

/ˌfwaː'ɡrɑː/
foie gras

焗蜗牛

/eskar'ɡɒ/
escargot

菲力牛排

/'fɪlɪt steɪk/
fillet steak

西冷牛排

/'sɜːlɔɪn steɪk/
sirloin steak

肋眼牛排

/'rɪb ˌaɪ steɪk/
rib eye steak

T骨牛排

/tiː bəun steɪk/
T-bone steak

胡椒子酱

/'pepəkɔːn sɔːs/
peppercorn sauce

酱汁

/sɔːs/
sauce

荷兰酱

/ˌhɒlən'deɪz/
hollandaise

一成熟

/'veri reə(r)/
very rare

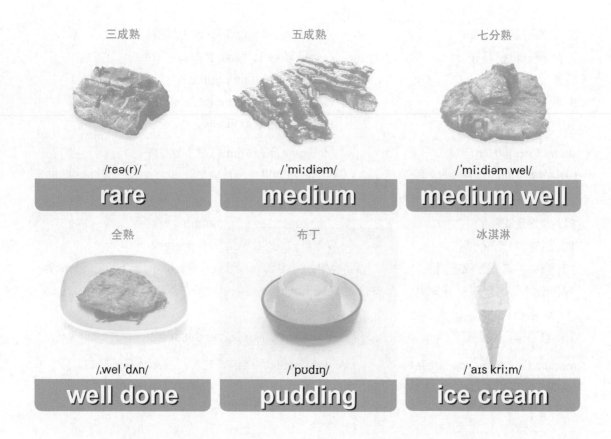

Functional Expressions 实用表达

请流利地朗读以下词组,并做到中英互译。

1. 三道菜的正餐包括开胃菜、主菜和甜点。
2. 菜品是上完一道再上下一道。
3. 六道菜的晚餐菜单包括冷盘、汤、开胃菜、沙拉、主菜和甜点。
4. 菜单是按用餐顺序来设计的。
5. 许多客人会把沙拉作为第一道菜。

The three-course meal includes an appetizer, main course, and dessert.
Dishes are served one after the other.
A 6-course dinner menu includes an hors d'oeuvre, soup, appetizer, salad, main course, and dessert.
The menu is planned in the order of a meal.
Many guests will start with the salad as the first course.

three-course /θriː kɔːs/ 三道菜
appetizer /ˈæpɪtaɪzə(r)/ **n.** 开胃菜,头盘
dessert /dɪˈzɜːt/ **n.** 甜点
after /ˈɑːftə(r)/ **prep.** 在……之后
hors d'oeuvre /ɔːˈdɜːvr/ (法)开胃小菜,冷盘
order /ˈɔːdə(r)/ **n.** 顺序
first course /fɜːst kɔːs/ 第一道菜

course /kɔːs/ **n.** 一道菜
main course /ˌmeɪn ˈkɔːs/ 主菜
served /sɜːvd/ **v.** 供应(食物)
other /ˈʌðə(r)/ **pron.** 另一个;其他
soup /suːp/ **n.** 汤
guests /ɡests/ **n.** 顾客(guest 的复数)

6. 您的头盘要什么？	What would you like for starter?
7. 我要一份鹅肝酱。	I would like to have the foie gras.
8. 那么主菜呢？	And for the main course?
9. 我要一份菲力牛排。	I would like to have the fillet steak.
10. 您的牛排要几成熟？	How would you like your steak?

starter /ˈstɑːtə(r)/ **n**. 头盘　　　　　　　foie gras /ˌfwɑː ˈɡrɑː/（法）鹅肝酱
fillet steak /ˈfɪlɪt steɪk/ 菲力牛排　　　　steak /steɪk/ **n**. 牛排

11. 我要五成熟的。	I'd like it medium, please.
12. 你喜欢什么配菜？	What side dish do you prefer?
13. 您的牛排配什么酱汁？	What sauce would you like to go with your steak?
14. 我们有花椒酱、芥末酱、卤肉汁和番茄酱。	We have peppercorn sauce, mustard, gravy, and ketchup.
15. 您要来点甜点吗？	Would you like some dessert?

medium /ˈmiːdiəm/ **adj**. 五成熟的　　　side dish /ˈsaɪd dɪʃ/ 副菜，配菜　　sauce /sɔːs/ **n**. 酱汁
peppercorn sauce /ˈpepəkɔːn sɔːs/ 胡椒子酱　　　　　　　　　　　　mustard /ˈmʌstəd/ **n**. 芥末酱
gravy /ˈɡreɪvi/ **n**. 肉汁，卤肉汁　　　　ketchup /ˈketʃəp/ **n**. 番茄酱

16. 至于甜点，我要奶酪拼盘。	For dessert, I'd like to have the cheese platter.
17. 这是你的最后一道菜。	This is your complete course.
18. 味道如何？	How does it taste?
19. 我还能为您做些什么吗？	Is there anything else I can do for you?
20. 请慢用。	Please enjoy your meal.

platter /ˈplætə(r)/ **n**. 大浅盘，大平盘　　　cheese platter /tʃiːz ˈplætə(r)/ 奶酪拼盘
complete course /kəmˈpliːt kɔːs/ 最后一道菜　complete /kəmˈpliːt/ **adj**. 完整的；完全的
taste /teɪst/ **v**. 尝，吃　　　　　　　　　anything else /ˈeriθɪn els/ 别的东西

Task 任务　情景对话

请设计一段简单的对话，点一份三道菜的正餐。

三道菜：1. 头盘：恺撒沙拉　　　starter/appetizer: Caesar Salad
　　　　2. 主菜：菲力牛排　　　Main Course: Fillet Steak
　　　　3. 甜点：草莓冰淇淋　　Dessert: Strawberry Ice Cream

地点：Diamond Restaurant
人物：客人 Mr.White，资深餐厅服务员 Eric

Dialogues 对话

1. Table D'hote and A La Carte Menu 套餐及零点菜单

（初级厨师：Steve，行政副厨师长：Jackson）

Jackson:	In western restaurants, there are 2 types of food menus: the table d'hote and the a la carte menu. They are French words.
Steve:	Could the buffet menu be categorized as a form of table d'hote menu?
Jackson:	Yes, it's a form of table d'hote menu. A la carte menu offers choices of food and drinks in categories.
Steve:	Such as the luncheon and dinner menu.
Jackson:	That's correct. Usually, both table d'hote and a la carte menu has 4 parts: appetizers or starters, main course, desserts and drinks.
Steve:	I see.
Jackson:	Our menu is planned in the order of a meal. Now go through our English menu and learn new words.
Steve:	Yes, chef.

词汇

table d'hote /ˌtɑːbl 'dəʊt/ **adj**.（法）定餐的；套餐的
a la carte /ˌɑː lɑː 'kɑːt/ **adj**.（法）按菜单点菜
categorized /'kætəgəraɪzd/ **v**. 分类（categorize 的过去式）
form /fɔːm/ **n**. 形式
choices /'tʃɔɪsɪs/ **n**. 选择（choice 的复数）
categories /'kætəgərɪz/ **n**. 类别（category 的复数）；**v**. 分类
luncheon /'lʌntʃən/ **n**. 正式午宴
usually /'juːʒəli/ **adv**. 通常
appetizer /'æpɪtaɪzə/ **n**. 开胃菜，头盘
main course /ˌmeɪn 'kɔːs/ 主菜
dessert /dɪ'zɜːt/ **n**. 甜点
drinks /drɪŋks/ **n**. 饮料，饮品（drink 复数）
planned /plænd/ **v**. 打算（plan 的过去分词）；设计
go through /gəʊ θruː/ 通读；查阅；检查

中文

Jackson： 在西餐厅，菜单有两种：套餐菜单和零点菜单。它们是法语词。
Steve： 自助餐菜单可以归类为套餐菜单吗？
Jackson： 可以，它是套餐菜单的一种。零点菜单提供各种食物和饮料。
Steve： 例如午餐和晚餐菜单。
Jackson： 对。通常情况下，套餐和零点菜单包含四个部分：开胃菜或头盘，主菜、甜点和饮料。
Steve： 我懂了。
Jackson： 我们的菜单是按用餐顺序设计的。现在查阅我们的英文菜单并学习新的词汇。
Steve： 好的，厨师长。

注释

（1）table d'hote：table d'hote 是法语词汇，翻译为定制菜单或者套餐菜单。原意是主人餐桌提供的食物（food from the hosts' table）。西餐的套餐菜单包含了完整的餐序，每一道餐序可有数个选项，以固定的价格任客人取食。套餐菜单上的食物准备起来更简单，不需要很大的厨房空间及复杂的厨房设备。

（2）a la carte：a la carte 翻译为零点菜单、按菜单点菜或者逐道点菜菜单。法语词汇里"a la"意为"按……式的"，"carte"意为卡片、菜单。零点菜单指客人从菜单列表中自由选择菜式，菜单上每道菜都有单独的价格。零点形式的菜单更为流行。

（3）其他菜单。du jour menu（/du'ʒuə/）今日特色菜单，wine menu 酒单（或者称为"drink list"），dessert menu 甜点菜单。

2. Taking a 3-course Meal Order 点一份三道菜的正餐

（客人：Bella，资深服务员：Eric）

Eric: Good evening, madam. Here is our menu and the drink list. What aperitif would you like before your meal?
Bella: A negroni, please.
Eric: I'll be back with your negroni and take your order later. Just a moment.
Here's your aperitif. What would you like for appetizer?
Bella: A smoked salmon, please.
Eric: For your main course, may I recommend today's special? It's also our signature dish: bone-in rib-eye steak with peppercorn sauce.
Bella: I'd love to try it.
Eric: How would you like the steak?
Bella: Medium, please.
Eric: Would you like to have some dessert?
Bella: The chocolate strawberry crepes, please.
Eric: Alright. Let me confirm your order. Smoked salmon for starter, the bone-in rib-eye steak with peppercorn sauce for main course. And follow with chocolate strawberry crepes for dessert, is that right?
Bella: That's right.
Eric: Just a moment, madam.

词汇

madam /ˈmædəm/ n. 女士，夫人
drink list /drɪŋk lɪst/ 酒水单，饮料单
aperitif /əˌperəˈtiːf/ n. 餐前酒
negroni /nɪˈɡrəʊnɪ/ n. 内格罗尼酒（一种鸡尾酒，由金巴利、金酒及甜味美思混合而成）
take order /teɪk ɔːdə(r)/ 点酒水、点菜
aperitif /əˌperəˈtiːf/ n. 餐前酒，开胃酒
appetizer /ˈæpɪtaɪzə(r)/ n. 开胃菜，头盘
smoked /sməʊkt/ adj. 烟熏的
salmon /ˈsæmən/ n. 三文鱼，鲑鱼
main course /ˌmeɪn ˈkɔːs/ 主菜
recommend /ˌrekəˈmend/ v. 推荐
today's special /təˈdeɪs ˈspeʃl/ 今日特餐，今日特价商品
signature dish /ˈsɪɡnətʃə(r) dɪʃ/ 招牌菜，特色菜
signature /ˈsɪɡnətʃə(r)/ n. 明显特征；特色，识别标志
bone-in /bəʊn in/ 带骨的
rib eye steak /ˈrɪb ˌaɪ steɪk/ 肋眼牛排
peppercorn /ˈpepəkɔːn/ n. 胡椒粒
medium /ˈmiːdiəm/ adj. 半生熟的，五成熟
chocolate /ˈtʃɒklət/ n. 巧克力
strawberry /ˈstrɔːbəri/ n. 草莓
crepe /kreɪp/ n. 煎薄饼，可丽饼
confirm /kənˈfɜːm/ v. 确认；确定
starter /ˈstɑːtə(r)/ n. 头盘，开胃品

中文

Eric： 晚上好，女士。这是我们的菜单和酒水单。您要来点餐前酒吗？
Bella： 一杯内格罗尼酒。
Eric： 我这就去拿您的内格罗尼酒，稍后回来再为您点菜。请稍等。这是您的餐前酒。您要什么头盘？
Bella： 一份烟熏三文鱼。
Eric： 关于您的主菜，我能推荐今天的特色菜吗？这也是我们的招牌菜：带骨肋眼牛排配胡椒酱。
Bella： 我很乐意尝试。
Eric： 您的牛排要几成熟？
Bella： 五成熟。
Eric： 你要来些甜点吗？
Bella： 一份巧克力草莓薄饼。
Eric： 好的。我和您确认您一下，开胃菜是烟熏三文鱼，主菜是带骨肋眼牛排配胡椒酱。然后甜点是巧克力草莓薄饼，对吗？
Bella： 是的。
Eric： 女士，请稍等。

注释

（1）点饮料时，大部分情况下直接使用数词即可。比如：

　　a cider 一杯苹果酒
　　a coke 一份可乐
　　a campari 一杯金巴利酒
　　a Long Island iced tea 一杯长岛冰茶

饮料的量词使用与容器有关。有些鸡尾酒装在"shot glass"（烈酒杯，一种矮脚杯）里，因此若要确切地表达"一杯鸡尾酒"，则是：

　　a shot of cocktail

使用"shot glass"的还有：

　　a shot of whisky 一杯威士忌
　　a shot of vodka 一杯伏特加
　　2 shots of gin 一杯双份金酒

葡萄酒的传统表达依然是：

　　a glass of white wine 一杯白葡萄酒
　　a glass of champagne 一杯香槟酒
　　2 glasses of red wine 两杯红葡萄酒

（2）在菜单里，大份量的"bone-in rib eye steak"也会被称为"tomahawk"（/ˈtɒməhɔːk/，战斧牛排）。在一些英语国家里，肉类食物所使用的重量单位为盎司（ounce，缩写为 oz；1 盎司约为 28.3 克），因此英文菜单上的"32oz tomahawk steak"约为 900 克，可供多人食用。

（3）餐厅使用零点菜单（a la carte menu）意味着客人有点菜数量的自由，若不需要某一道餐序，可以说：

I'll skip the starter. 我不点头盘

I don't want dessert. 我不想吃甜点。

You don't like any dessert, do you? 你不喜欢吃甜点，是吗？

（4）中文使用数字来描述牛排的熟度，而英语里则是使用形容词来表达。例如中文的"五成熟"对应英文的"medium"（中等熟度的）；中文的"一成熟"对应英文的"very rare"（非常生的肉）。英语里肉类食物的熟度词汇为：

全熟　　well done

七成熟　medium well (medium to well done)

五成熟　medium

三成熟　rare

一成熟　very rare

（5）"招牌菜"的两个常见翻译：① signature dish；② house specialty。"signature"在这里有"代表性的"及"鲜明特色"之意。

（6）在点菜时，菜单上注明有数种酱汁、佐餐酒和蔬菜以供客人选择，此时可使用如下句型：

您的牛排要配什么酱料？

What sauce would you like to go with your steak?

您的牛排要配什么佐餐酒？

What wine would you like to go with your steak?

您的牛排要配什么蔬菜？

What vegetable would you like to go with your steak?

在表达"我要……"时，使用如下句型：

I'd like to have …

（7）几种常见的西餐酱汁：

peppercorn sauce /ˈpepəkɔːn sɔːs/ 胡椒子酱

hollandaise /ˌhɒləndeɪz/ 荷兰酱

mayonnaise /ˌmeɪəˈneɪz/ 蛋黄酱　　ketchup /ˈketʃəp/ 番茄酱

mustard /ˈmʌstəd/ 芥末酱　　　　gravy /ˈɡreɪvi/ 肉汁，卤肉酱

（8）三道菜的西餐正餐。

一份三道菜的西餐正餐包括头盘、主菜和甜点。

Service procedures of ordering a 3-course meal. 点3道菜正餐的服务流程。

第一道菜（the first course）

询问是否点头盘：

Would you like to order now?

What would you like for starter?

What would you like to start with?

第二道菜（the second course）

询问是否点主菜：

And for the main course?

What would you like for your main course?

What would you like to follow with?

主菜所涉及的选项：

询问牛排的熟度

How would you like your steak?

询问配菜

What side dish do you prefer?

询问酱汁的种类

What sauce would you like to go with it?

第三道菜（the third course）

询问是否点甜点：

Would you like some dessert?

Would you like to look at the dessert menu?

Anything for dessert?

* 客人或许在吃完主餐之后再决定是否点甜点。

Would you like to have a look at the dessert menu?

Any tea or coffee?

Would you like it now or later?

与客人确认所点菜品

You've ordered...for starter, a...for main course, and follow with...for dessert, is that right?

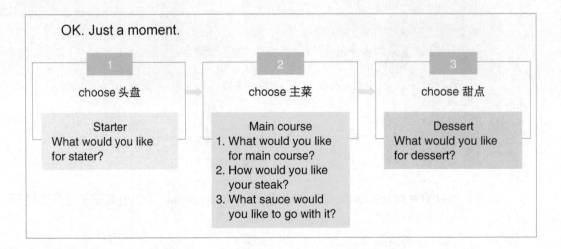

3. Western Menu Translation 西餐菜单翻译

（初级厨师：Steve，行政副厨师长：Jackson）

Jackson: In general, there are three types of cooking methods: the dry heat cooking, moist heat cooking, and combination cooking.
Steve: As far as I know, there are many techniques to transfer heat to foods.
Jackson: Yes! You might come across some cooking terms that are unfamiliar to you.
Steve: I cannot fully understand our menu and those narrations underneath each dish.
Jackson: Here is an assignment for you: translate this menu to Chinese.
Steve: Yes, chef. Right away.

词汇

in general /ɪn ˈdʒenrəl/ 总之，通常
dry heat /draɪ hiːt/ 干热
heat /hiːt/ **n.** 高温；热度
moist /mɔɪst/ **adj.** 潮湿的；湿润的
combination /ˌkɒmbɪˈneɪʃn/ **n.** 组合，混合
technique /tekˈniːk/ **n.** 技巧，技术
transfer /trænsˈfɜː(r)/ **v.** 转移
come across /kʌm əˈkrɒs/ **v.** 遇见；无意中发现
term /tɜːm/ **n.** 术语；措辞；词语
unfamiliar /ˌʌnfəˈmɪliə(r)/ **adj.** 不熟悉的，陌生的
fully understand /ˈfʊli ˌʌndə(r)ˈstænd/ 完全理解
narration /nəˈreɪʃn/ **n.** 讲述，故事；解说
underneath /ˌʌndə(r)ˈniːθ/ **prep.** 在……底下
assignment /əˈsaɪnmənt/ **n.** 任务，作业
translate /trænsˈleɪt/ **v.** 翻译

中文

Jackson: 总的来说，有三种烹饪方法：干热法、湿热法和混合法。
Steve: 据我所知，有很多烹饪技法可以将热量传递给食物。
Jackson: 对！你会遇到一些不熟悉的烹饪术语。
Steve: 我看不太懂我们的菜单和每道菜下面的解说。
Jackson: 这里有一项任务给你：把这份菜单翻译成中文。
Steve: 好的，厨师长。马上做。

> **注释**
>
> （1）厨房日常对话涉及大量烹饪专业词汇、菜品原材料及一些常识性的英语单词，还包含一些外来语，如法语、意大利语及日语等。比如，"铁板烧"的英文词汇采用日语音译词"teppanyaki"；"焗蜗牛"的英文词汇直接采用法语"escargot"；意大利不同的面型词汇本身即是意大利语。随身常备翻译词典是学好烹饪英语的关键。
>
> （2）一份正式的西餐菜单会注明食材及烹饪方法。如：
>
>
>
> JUicy MIxed Fruit Salad
>
> Pineapple, melon and berries. It's the pineapple syrup combined with the juice of the orange, delicious and refreshing.
>
>
>
> 西餐菜名通常包括以下三个部分：
>
>   1）菜的名称（the name of the dish）
>
>   2）食材（the ingredients）
>
>   3）描述、解说（description/narration）
>
> 好的菜式描述会增强食客的选择意愿，以此增加餐厅的收入。
>
> （3）常见的干热法。
>
>   煎 frying/sauteing   烧烤 grilling/barbequing
>
>   烤 baking/roasting   炸 deep-frying
>
> （4）常见的湿热法。
>
>   蒸 steaming   水煮 boiling/poaching
>
>   慢炖 simmering   焯水 blanching
>
> （5）常见的混合法。
>
>   炖、焖（先炒后炖）stewing/braising

练习：西餐菜单翻译

请同学们准备好翻译工具——字典以及可以连接互联网的电脑或手机，做完以下西餐菜单翻译练习。同时请记得查询图片以增强对菜名的理解。

	Western Menu Translation 西餐菜单翻译练习				
	Starter				
1	juicy fruit salad		2	seafood salad	
3	caesar salad		4	chef's salad	
5	nicoise salad		6	tuna salad	
7	caviar（源自法语）		8	foie gras（源自法语）	
9	cream of mushroom soup		10	beef consommé	
11	mackerel cake		12	poached leek salad	

（续）

	Western Menu Translation 西餐菜单翻译练习				
	Starter				
13	cocktail shrimp		14	minestrone soup	
15	lobster bisque		16	calamari（源自意大利语）	
17	seafood chowder		18	clam casino	
19	french onion soup		20	gazpacho（源自西班牙语）	
21	grand aioli		22	bouillabaisse	
	Main Course				
23	chicken cordon bleu		24	chicken stew with potatoes	
25	beef teppanyaki		26	roast turkey with apple cider brine	
27	braised chicken with red wine		28	deep-fried chicken cutlet	
29	roast breast of chicken		30	char-grilled octopus	
31	risotto（源自意大利语）		32	pesto chicken with pickled cucumber	
33	hand chopped beef ravioli		34	seafood spaghetti（源自意大利语）	
35	macaroni with seafood		36	skillet lasagna	
37	dijon meatloaf		38	cast iron roast chicken with fennel and carrots	
39	scrambled egg		40	curry chicken	
41	chicken in red wine vinegar		42	calves liver, sherry vinegar and caper sauce	
43	slow-grilled leg of lamb		44	boneless prime rib roast with cream sauce	
45	grilled rib-eye steak		46	smoked spare ribs with honey	
47	roast sirloin steak with red wine sauce		48	stew beef	
49	t-bone steak		50	fillet steak	
51	deep-fried spare ribs with garlic		52	grilled lamb chops	
53	grilled garlic and black pepper shrimp		54	red snapper with citrus and fennel salad	
55	fried whole fish with tomatillo sauce		56	roast lamb chops in cheese and red wine	
57	wagyu beef rib eye		58	lamb shanks with pomegranate and walnuts	

（续）

<table>
<tr><td colspan="4" align="center">**Western Menu Translation**
西餐菜单翻译练习</td></tr>
<tr><td colspan="4" align="center">Main Course</td></tr>
<tr><td>59</td><td>green curry pork tenderloin</td><td>60</td><td>clam toasts with pancetta</td></tr>
<tr><td>61</td><td>curry beef</td><td>62</td><td>fried eggs with ham</td></tr>
<tr><td>63</td><td>beef Wellington</td><td>64</td><td>vegetarian pizza</td></tr>
<tr><td>65</td><td>mediterranean seabass</td><td>66</td><td>seafood pizza</td></tr>
<tr><td>67</td><td>braised ox tongue</td><td>68</td><td>grilled pork belly</td></tr>
<tr><td>69</td><td>bbq chicken leg</td><td>70</td><td>bbq lobster tail</td></tr>
<tr><td colspan="4" align="center">Desserts</td></tr>
<tr><td>71</td><td>rye bread</td><td>72</td><td>waffle</td></tr>
<tr><td>73</td><td>doughnut</td><td>74</td><td>strawberry cake</td></tr>
<tr><td>75</td><td>flan</td><td>76</td><td>crème brûlée（源自法语）</td></tr>
<tr><td>77</td><td>tiramisu</td><td>78</td><td>cheesecake</td></tr>
<tr><td>79</td><td>cream puff</td><td>80</td><td>soufflé（源自法语）</td></tr>
<tr><td>81</td><td>éclair（源自法语）</td><td>82</td><td>croissant</td></tr>
<tr><td>83</td><td>pancake</td><td>84</td><td>chocolate mousse</td></tr>
<tr><td>85</td><td>biscuit</td><td>86</td><td>egg custard tart</td></tr>
<tr><td>87</td><td>onion loaf</td><td>88</td><td>raspberry pavlova</td></tr>
<tr><td>89</td><td>cranberry and raisin scones</td><td>90</td><td>walnut brownies</td></tr>
<tr><td>91</td><td>pretzel</td><td>92</td><td>sponge cake</td></tr>
<tr><td>93</td><td>chocolate fudge</td><td>94</td><td>swiss roll</td></tr>
<tr><td>95</td><td>rum and raisin ice cream</td><td>96</td><td>panna cotta with pineapple</td></tr>
<tr><td>97</td><td>whole wheat bread</td><td>98</td><td>napoleon pastry</td></tr>
<tr><td>99</td><td>coconut rice pudding</td><td>100</td><td>white chocolate and raspberry trifle</td></tr>
</table>

职业提示

全球素养，兼收并蓄

习近平总书记在纪念孔子诞辰 2565 周年国际学术研讨会上说："本国本民族要珍惜和维护自己的思想文化，也要承认和尊重别国别民族的思想文化。"作为一名厨师，我们不仅要精通中式菜肴的烹饪技法，也要了解西餐的历史、文化及礼仪，促进我们与世界各族人民的交流互鉴。

Unit 10 Menu of Chinese Food
中餐菜单

与西餐的分餐制不同，中餐属于聚餐制。中餐点菜时往往需同别人商量，菜单上较少文字阐释，中文菜名简洁。中国烹饪文化历史悠久、底蕴深厚，中餐菜名在英译时既要参考西餐菜名的叫法，还要在保留其文化寓意的前提下便于外宾理解，这是一个不小的挑战。在掌握中餐菜单的翻译技巧之前，先了解一些经典的中餐菜品的英文译名。

Learning Objectives 学习目标

❋ Memorize basic terms of food cooking.
   记住烹调的基本术语。
❋ Know some classic Chinese dishes.
   了解一些经典的中餐菜品。

扫码看视频

Basic Terms 基础词汇

菜肴，烹饪术 cuisine /kwɪˈziːn/	风格的 style /staɪl/	八大菜系 8 major cuisines /eɪt ˈmeɪdʒə(r) kwɪˈziːns/
颜色 color /kʌlə(r)/	香味 aroma /əˈrəʊmə/	滋味 flavor /fleɪvə/
鲜味 umami /uːˈmɑːmi/	摆盘 plating /ˈpleɪtɪŋ/	瓜雕 melon carving /ˈmelən ˈkɑːvɪŋ/
新鲜的 fresh /freʃ/	咸味的 savoury /ˈseɪvəri/	辣的 spicy /spaɪsi/
麻辣的 numbing hot /nʌmɪŋ hɒt/	清淡的 light /laɪt/	浓重的 heavy /ˈhevi/
可食用的 edible /ˈedəb(ə)l/	本地的 local /ˈləʊk(ə)l/	地道的 authentic /ɔːˈθentɪk/
食材 food material /fuːd məˈtɪəriəl/	野生食材 wild ingredients /waɪld ɪnˈɡriːdiənts/	具有代表性的 representative /ˌreprɪˈzentətɪv/
传统的 traditional /trəˈdɪʃ(ə)n(ə)l/	创新的 creative /kriˈeɪtɪv/	无国界料理 fusion /ˈfjuːʒn/

Functional Expressions 实用表达

请流利地朗读以下句子，并做到中英互译。

1. "色、香、味"是中餐的三大要素。 "Color, aromatic flavor, taste" are the three essential factors of Chinese food.
2. 中国有很多种菜系，但最具影响力的是"中国八大菜系"。 There are many styles of cooking in China, but the 8 Major Chinese Cuisines are the most influential.
3. 粤菜在世界各地都很受欢迎。 The Cantonese cuisine is popular around the world.
4. 粤菜讲究原材料的原味及鲜嫩。 Cantonese cuisine emphasizes the original taste and freshness of the ingredients.
5. 川菜的特点是辛辣。 Sichuan cuisine is characterized by its hot and pungent flavor.

aromatic /ˌærəˈmætɪk/ *adj*. 芳香的
essential /ɪˈsenʃl/ *adj*. 基本的，必要的
8 major Chinese cuisines /eɪt ˈmeɪdʒə(r) ˌtʃaɪˈniːz kwɪˈziːns/ 中国八大菜系
influential /ˌɪnfluˈenʃl/ *adj*. 有影响力的
Cantonese cuisine /ˌkæntəˈniːz kwɪˈziːn/ 粤菜
original /əˈrɪdʒənl/ *adj*. 原始的；原来的
freshness /ˈfreʃnəs/ *n*. 新，新鲜
be characterized by /bi ˈkærəktəraɪzd baɪ/ 有……特点
pungent /ˈpʌndʒənt/ *adj*. 刺鼻的；辛辣的

flavor /ˈfleɪvə(r)/ *n*. 味道，口味
factor /ˈfæktə(r)/ *n*. 因素，要素

Cantonese /ˌkæntəˈniːz/ *adj*. 广东人的
emphasize /ˈemfəsaɪz/ *v*. 着重于；强调
original taste /əˈrɪdʒənl teɪst/ 原汁原味
ingredient /ɪnˈɡriːdiənt/ *n*. 原料
hot /hɒt/ *adj*. 辣的

6. 鲁菜以咸鲜为主，有较多的海鲜菜式。 Shandong cuisine is mainly salty and fresh, having more of seafood dishes.
7. 淮扬菜清淡、鲜嫩且精致。 Huaiyang cuisine is light, fresh, and delicate.
8. 广东人喜欢炖汤。 Cantonese like to boil soups.
9. 食材必须是新鲜的。 The ingredients must be fresh.
10. 它使用了大量的海鲜、河鲜和当季的食材。 It uses a lot of fresh seafood, river food, and ingredients of the season.

mainly /ˈmeɪnli/ *adv*. 主要地
seafood /ˈsiːfuːd/ *n*. 海鲜
delicate /ˈdelɪkət/ *adj*. 清淡可口；精美的
river food /ˈrɪvə(r) fuːd/ 河鲜；在河里生长的食物

salty /ˈsɔːlti/ *adj*. 咸的
light /laɪt/ *adj*. 清淡的
boil soup /bɔɪl suːp/ 煮汤
of the season /ɒv ðə ˈsiːzn/ 当季的

fresh /freʃ/ *adj*. 新鲜的

11. 菜肴口味清淡、新鲜，保留了食材的天然味道。	The dish tastes mild and fresh, preserving the natural taste of food materials.
12. 它的味道非常浓郁。	It has a rich flavor.
13. 这道菜看起来非常吸引人。	This dish is visually appealing.
14. 这道菜色香味俱全（味道鲜美、香气浓郁、摆盘精致）。	The dish is tasty and richly aromatic with fine presentation.

tastes /teɪsts/ **v.** 吃，尝（taste 的第三人称单数）
preserving /prɪˈzɜːvɪŋ/ **v.** 保留，维持
materials /məˈtɪəriəlz/ **n.** 材料
visually appealing /ˈvɪʒuəli əˈpiːlɪŋ/ 视觉上吸引人的
fine presentation /faɪn preznˈteɪʃn/ 摆盘精致

mild /maɪld/ **adj.** 温和的；淡味的
natural taste /ˈnætʃrəl teɪst/ 天然的味道
rich /rɪtʃ/ **adj.** 浓郁的；油腻的
richly /ˈrɪtʃli/ **adv.** （味道）浓郁地

15. 人们围坐在餐桌旁，分享不同的菜肴。	People sit around the table and share different dishes.
16. 菜品摆放在桌子的中央。	Dishes are placed in the center of the table.
17. 对中国人来说，食物也是良药。	For Chinese, food is also medicine.
18. 选择肉类、蔬菜、米饭、饮料和甜点来做一顿饭。	Choose meats, vegetables, rice dishes, beverages, and dessert to complete a meal.

placed /pleɪst/ **v.** 放置（place 的过去分词形式）
beverages /ˈbevərɪdʒz/ **n.** 饮料

medicine /ˈmedɪsn/ **n.** 药

Task 任务　菜系描述

请简单描述中国四大菜系的特点。
四大菜系：1. 粤菜 Cantonese/Guangdong Cuisine: light, mild, fresh, natural, original
　　　　　2. 川菜 Sichuan Cuisine: spicy, numbing hot, Sichuan pepper
　　　　　3. 淮扬菜 Huaiyang Cuisine: umami, fresh and crisp, original, fine presentation
　　　　　4. 鲁菜 Shandong Cuisine: umami, salty, more stewing and oil

Dialogues 对话

1. Beijing Roast Duck 北京烤鸭

（初级厨师：Steve，中餐厨师长：David）

David: Beijing roast duck is also known as Peking roast duck, as Peking is the old spelling. This is one of the most celebrated dishes in China.

Steve: It's a famous and time-honored dish.

词汇

roast /rəʊst/ **adj.** 烘烤的，烤制的
duck /dʌk/ **n.** 鸭肉
Peking /ˌpiːˈkɪŋ/ **n.** 北京的英文旧称（等于 Beijing）
spelling /ˈspelɪŋ/ **n.** 拼写

David: Right. Guests will be watching how the duck is carved by the chef on table-side.
Steve: It's a lot of fun for customers.
David: Uh-huh. I will show you how to slice the duck and wrap the duck meat in a piece of pancake.
Steve: Yes, chef.

celebrated /ˈselɪbreɪtɪd/ adj. 著名的；有名望的
time-honored /ˈtaɪmɒnəd/ adj. 历史悠久的；因古老而受到尊重的
carved /kɑːvd/ v. 雕刻，切（carve 的过去式和过去分词）
uh-huh /ʌ ˈhʌ/ int. 嗯哼（表示同意、理解或给出肯定答案）
slice /slaɪs/ v. 将……切成薄片
wrap /ræp/ v. 包，裹
a piece of /ə piːs ɒv/ 一片，一块
pancake /ˈpænkeɪk/ n. 煎薄饼

中文

David：北京烤鸭也被称为北平烤鸭，因为北平是北京的旧称。这是中国最著名的菜肴之一。
Steve：这是一道著名且历史悠久的菜。
David：对。客人会在餐桌旁观看厨师如何切割鸭子。
Steve：对客人来说很有趣。
David：嗯。我给你展示如何将鸭子切成薄片并将鸭子肉包进薄饼里。
Steve：好的，厨师长。

注释

（1）"Peking"一词自 17 世纪由法国传教士传到欧洲后，西方国家以此称呼北京城。新中国成立以后采用了新的拼音方法，从 1979 年开始，中国官方媒体正式将"Peking"更名为"Beijing"。

（2）用英语描述北京烤鸭的吃法。

在客人面前将鸭子切成薄片	slice the duck in front of guests
在你的手上放一块饼皮	lay a pancake in your hand
用饼皮包住鸭肉	wrap the duck meat in a piece of pancake
将葱丝和黄瓜蘸酱	dip a few scallion and cucumber in the sauce
把酱涂在饼皮上	spread some sauce over the pancake
把它们放在饼皮上	lay them in the pancake
把饼皮卷起来	fold the pancake
咬一口	take a bite
鸭皮酥脆	the skin is crispy
鸭肉多汁	the duck meat is juicy

2. The Order of Serving Chinese Food 中餐的上菜顺序

（初级厨师：Steve，资深服务员：Eric）

Eric: In a Chinese menu, there are meat dishes, vegetable dishes, rice dishes, noodle dishes, beverages, and desserts.

Steve: Can you tell me what's the order of serving Chinese food?

Eric: Tea usually is served while customers read the menu and decide what to order.

Steve: It's reasonable.

Eric: Cold dishes come first, then the beverage or wine. Follow with soup. Hot dishes such as fried dishes and stewed dishes are served later on.

Steve: Dessert is the last dish.

Eric: Yes. Unlike in the West, hot dishes will arrive in the table at different times, depending on the chef's pace.

Steve: The whole meal will be served at one time if all dishes are ready.

Eric: That's true.

词汇

dishes /ˈdɪʃɪz/ **n.** 菜肴；餐具（dish 的复数）
beverage /ˈbevərɪdʒ/ **n.** 饮料
order /ˈɔːdə(r)/ **n.** 顺序
serving /ˈsɜːvɪŋ/ **v.** 为……服务（serve 的现在分词）
usually /ˈjuːʒəli/ **adv.** 通常
customer /ˈkʌstəmə(r)/ **n.** 顾客
reasonable /ˈriːznəbl/ **adj.** 合理的，公道的
fried dishes /fraɪd ˈdɪʃɪz/ 炒菜
fried /fraɪd/ **adj.** 炒，煎的；油炸的
stewed dishes /ˈstjuːd ˈdɪʃɪz/ 炖菜
stewed /stjuːd/ **adj.** 慢慢炖的
last /lɑːst/ **adj.** 最后的
unlike /ˌʌnˈlaɪk/ **prep.** 不像
depending on /dɪˈpendɪŋ ɒn/ 取决于
pace /peɪs/ **n.** 步伐；速度

中文

Eric: 在中文菜单里，有肉菜、蔬菜、米饭、面条、饮料和甜点。

Steve: 你能告诉我中餐的上菜顺序是什么吗？

Eric: 通常在顾客阅读菜单并决定点什么的时候上茶。

Steve: 这很合理。

Eric: 首先是冷碟，然后是饮料或者酒。接着是汤。热菜比如炒菜和炖菜等再晚一些上菜。

Steve: 甜品是最后的餐序。

Eric: 对。与西方不同，根据厨师的节奏，热菜会在不同的时间送上餐桌。

Steve: 如果所有菜都准备好了，整餐会一次上齐。

Eric: 是的。

注释

（1）中西餐上菜顺序。

1）西餐的上菜顺序（以六道菜正餐为例）。

冷盘（hors d'oeuvre）——汤（soup）——头盘（appetizer）——沙拉（salad）——主菜（main course）——甜点（dessert）

2）中餐的上菜顺序。

冷盘（cold dishes）——酒水（beverage and wine）——汤（soup）——热菜（hot dishes）——主食（staple food）——水果和甜点（fruits and dessert）

（2）中餐的冷菜（cold dishes）也叫冷盘、凉菜，讲究装盘。通常为凉拌菜、腌渍小菜，在英译时通常直接翻译食材和调料，比如：

手拍黄瓜	smashed cucumbers with sauce
野山椒凤爪	chicken feet with pickled peppers
红油笋干	bamboo shoots in red chili oil

（3）中餐的主食（staple food）有米饭、面条、粥、杂粮饭等，相关词汇如下：

白米饭 steamed rice	杂粮饭 mixed grain rice	小米粥 millet congee
面条 noodles	米粉 rice noodles	煎饼 pancake
馒头 mantou(steamed bun)		

3. Chinese Menu Translation 中餐菜单翻译

（初级厨师：Steve，中餐厨师长：David）

David: There will be a 15-table year-end banquet this Friday evening. The menu has just been planned. Come on, take a look.

Steve: It's an upscale dinner menu. There are abalone, crab, lobster, and sea bass, etc.

David: The budget for each table is around 6,000 RMB.

Steve: Wow, super.

David: Here is a task for you: translate this menu into English.

Steve: Yes, chef.

词汇

there will be /ðeə(r) wɪl bi/ 将有（一般将来时）

15-table /ˌfɪfˈtiːn ˈteɪbl/ 15 桌

year-end /jɪə(r) end/ 年终的，年末的

banquet /ˈbæŋkwɪt/ **n**. 宴会，盛宴

planned /plænd/ **v**. 设计（plan 的过去分词）

upscale /ˌʌpˈskeɪl/ **adj**. 高消费阶层的，高档的

abalone /ˌæbəˈləʊni/ **n**. 鲍鱼

crab /kræb/ **n**. 螃蟹

lobster /ˈlɒbstə(r)/ **n**. 龙虾

sea bass /siː beɪs/ 石斑鱼

etc. /ɪtˈsetərə/ **abbr**. 等等 (et cetera)

budget /ˈbʌdʒɪt/ **n**. 预算

super /ˈsuːpə(r)/ **adj**. 顶呱呱，好极了

task /tɑːsk/ **n**. 任务，工作

中文

David：本周五晚上我们有一个 15 桌的年会宴席。菜单已经设计好了。过来看一看。
Steve：这是高档的晚餐菜单。有鲍鱼、螃蟹、龙虾和石斑鱼。
David：每桌的预算约为 6000 元人民币。
Steve：哇，好厉害。
David：给你布置一个任务：将这份菜单翻译成英文。
Steve：好的，厨师长。

注释

（1）"15-table"及"year-end"都是复合形容词。两个及两个以上的单词（名词、副词、动词等）使用连字符（-）可以合成一个复合形容词，以一个词进行简明扼要地表达。比如：

50 桌婚宴	fifty-table wedding banquet
刚做好的菜	newly-made dish
手擀面	hand-made noodles
手撕牛肉	hand-shredded beef
18 岁的男生	an eighteen-year-old boy

（2）宴席的常见种类。

商务宴会	business banquet	国宴	state banquet
生日宴会	birthday banquet	毕业宴会	graduation banquet
结婚宴会	wedding banquet	圣诞宴会	Christmas banquet
欢迎宴会	welcome banquet		
满月宴	full-moon-birth dinner		

注：国外是在宝宝出生前举行庆生会"baby shower"。

（3）在形容商品、食材、服务"高档、高级"时，可以使用的词汇有：

high-end, upscale, exclusive, upmarket

以下是常见的固定搭配表达：

高端服务	high-end service
高端炊具	high-end cookware
高档餐厅	high-end restaurants/fine dining restaurant
高端消费者	upscale consumers
优质食材	premium/superior ingredients

练习：中餐菜单翻译

将以下 18 道中餐菜式翻译为中文。需要说明的是，不同的餐馆在为菜式取名时皆采用不同的文字表达方式，以下菜式的中文名称没有统一的答案。

	Chinese Menu Translation 中餐菜单翻译练习	
1	assorted appetizers	
2	wild mushroom soup with black truffle	
3	steamed stuffed bamboo fungus	
4	fried sliced pork with chilli and garlic	
5	braised beef in brown sauce	
6	lemongrass roasted chicken	
7	braised shark's fin with crab meat	
8	fried shrimps with spiced salt	
9	sweet and sour mandarin fish	
10	stir-fried celery and peppers with octopus and pork	
11	steamed grouper fish with ginger and spring onion	
12	steamed scallops with vermicelli noodles and garlic	
13	abalone with Chinese cabbage and duck's feet	
14	steamed glutinous rice with Chinese preserved sausage wrapped in lotus leaf	
15	wok-fried seasonal vegetables	
16	pineapple bun	
17	red bean dessert soup	
18	seasonal fresh fruit platter	

职业提示

使命担当，文化自信

党的二十大报告首次将文明传播力、影响力与国际传播能力放在一起阐述，明确提出增强中华文明传播力影响力，坚守中华文化立场，讲好中国故事、传播好中国声音，展现可信、可爱、可敬的中国形象，推动中华文化更好走向世界。我们的祖国地大物博，几千年来形成了丰富的菜系，各地域、各民族的佳肴哺育了这片广袤大地上的华夏儿女。新时代的年轻人有责任继承优良的饮食文化传统，苦练烹饪基本功，创新性地融入现代文明，向世界展示我国饮食文化的繁荣和兴盛。

Part 4
Ingredients
食 材

Unit 11 Translation Techniques of Chinese Food 中餐菜名翻译技巧 / 116

Unit 12 Meats and Poultry 肉类 / 130

Unit 13 Aquatic Products 水产品 / 141

Unit 14 Food Tastes and Textures 食物的味道及口感 / 153

Unit 11 Translation Techniques of Chinese Food　中餐菜名翻译技巧

英文菜单涉及大量词汇，包括蔬果、肉类、海鲜、烹饪方式、刀工、烹饪器具、地名（甚至是文化）、方言或音译名等，有些菜单还会在菜名下方备注其起源、特殊的风味、食用方式、分量以及熟度。本单元我们学习如何将中文菜名翻译为英文。

本文归纳的中文菜单英文译名方法借鉴了 2008 年北京市外办和市民讲外语办公室联合出版的《美食译苑——中文菜单英文译法》一书。此书在北京举办奥运盛会、接待全球来宾之际对国内中文菜单的翻译进行了统一管理，是官方发布的权威中文菜名英文译法规范，具有重要指导作用。

Learning Objectives 学习目标

※ Learn principles and techniques of menu translating.
学习菜单翻译的原则和技巧。

※ Memorize the terms of cooking methods.
记住烹饪方式的词汇。

扫码看视频

Basic Terms 基础词汇

煎，炒
/fraɪ/
fry
pan-frled 用平底锅煎的

翻炒
/ˈstɜːfraɪ/
stir-fry

油炸
/ˌdiːp ˈfraɪ/
deep-fry

Unit 11　Translation Techniques of Chinese Food 中餐菜名翻译技巧

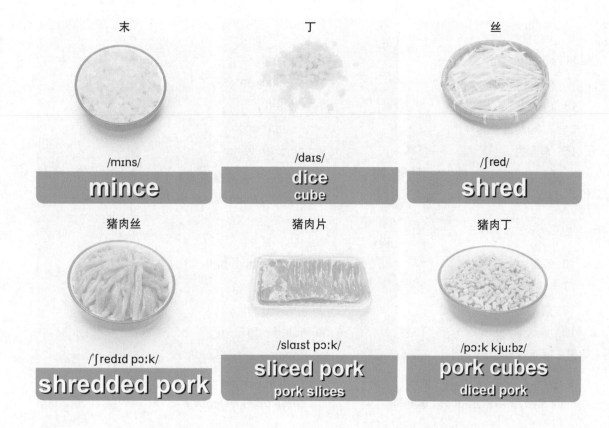

Functional Expressions 实用表达

流利地朗读以下菜名，并做到中英互译。

1. 炒猪肉		stir-fried pork
2. 炒肉片		wok-fried sliced pork
3. 炒肉丝		stir-fried shredded pork
4. 炒白菜		sauteed Chinese cabbage
5. 爆炒猪大肠		quick fried pork intestine

stir-fried /ˈstɜːfraɪ/ *adj.* 炒的　　stir /stɜː(r)/ *v.* 搅拌　　wok-fried 爆炒
fried /fraɪd/ *adj.* 油炸的；油煎的；油炒的　　wok /wɒk/ *n.* 锅（粤语）
sliced /slaɪst/ *adj.* （食物）已切成薄片的　　shredded /ˈʃredɪd/ *adj.* 撕碎的
sautéed /ˈsəʊteɪd/ *adj.* 煎的, 炒的（法语）　　quick fried /kwɪk fraɪd/ 快炒，爆炒

6. 炒蛋	scrambled egg
7. 煎蛋	fried egg
8. 煎牛肉	pan-fried beef
9. 蒸肉饼	steamed minced pork
10. 蒸鱼	steamed fish

scrambled /ˈskræmbld/ *adj.* 炒（蛋）　　　　　　pan /pæn/ *n.* 平底锅
pan-fried /ˈpæn fraɪd/ *adj.* 用平底锅煎的　　　　steamed /stiːmd/ *adj.* 蒸的

11. 炖牛肉	stewed beef
12. 焖鸡肉	braised chicken
13. 白灼虾	boiled shrimp
14. 白灼西蓝花	poached broccoli
15. 炸鸡翅	fried chicken wing

stewed /stjuːd/ ***adj***. 炖的（用文火慢慢煨炖）
boiled /bɔɪld/ ***adj***. 水煮的（沸水）
braised /breɪzd/ ***adj***. 焖的（先稍煎然后炖）
poached /pəʊtʃt/ ***adj***. 水煮的（高温但不沸腾）

16. 炸虾丸	deep-fried prawn ball
17. 烤鸡腿	roast chicken leg
18. 卤牛肉	marinated beef
19. 卤鸡蛋	marinated egg
20. 砂锅米粉	rice noodles in casserole

deep-fried /ˌdiːp ˈfraɪd/ ***adj***. 油炸的
marinated /ˈmærɪneɪtɪd/ ***adj***. 腌制的
roast /rəʊst/ ***adj***. 烤的
casserole /ˈkæsərəʊl/ ***n***. 砂锅，炖锅菜；***v***. 用焙盘焙

21. 瓦煲饭	clay pot rice
22. 汽锅鸡	steam pot chicken
23. 铁板牛肉	beef served on sizzling plate
24. 炭烤牛排	char-grilled steak
25. 真空慢煮牛排	sous vide beef

clay pot /kleɪ pɒt/ 瓦煲，黏土埚
steam pot /stiːm pɒt/ 蒸锅，汽锅
plate /pleɪt/ ***n***. 碟子
char-grilled /tʃɑ(r)grɪld/ ***adj***. 炭烤的 (char-grilled 同 chargrilled 和 charcoal grilled)
charcoal /tʃɑːkəʊl/ ***n***. 木炭
pot /pɒt/ ***n***. 壶，罐，锅
sizzling /ˈsɪzlɪŋ/ ***adj***. 极热的
sous vide /ˌsuːˈviːd/ 真空低温烹调（法语）

Task 任务　单词背诵

背诵烹饪方法词汇。

考核：同学们以组为单位，互相抽查词汇记忆。

Dialogues 对话

1. The Cooking Methods 烹饪方式

（初级厨师：Steve，中餐厨师长：David）

David: These are the basic cooking methods you need to know: frying, stir-frying, grilling, roasting, baking, boiling, poaching, steaming, simmering, stewing, and braising.

Steve: I'm unfamiliar with these cooking terms.

David: I understand. Some methods use dry heat, some use wet heat, and the other use the combination.

Steve: I see.

David: Some dishes may use more than one cooking method. But first of all, you need to memorize these terms.

Steve: Yes, Chef.

词汇

basic /ˈbeɪsɪk/ **adj.** 基本的
cooking method /ˈkʊkɪŋ ˈmeθəd/ 烹饪方法
frying /ˈfraɪɪŋ/ **n.** 油炸，油煎
stir-frying /ˈstɜː(r) fraɪɪŋ/ **n.** 翻炒
grilling /ˈɡrɪlɪŋ/ **n.** 烧烤，炙烤
roasting /ˈrəʊstɪŋ/ **n.**（食物）烘烤
baking /ˈbeɪkɪŋ/ **n.** 烘焙
boiling /ˈbɔɪlɪŋ/ **n.** 水煮（煮沸）
poaching /ˈpəʊtʃɪŋ/ **n.** 水煮（水温低一些）
simmering /ˈsɪmərɪŋ/ **n.** 煨，慢炖（文火）
stewing /ˈstjuːɪŋ/ **n.** 炖，焖
braising /ˈbreɪzɪŋ/ **n.** 焖（先煎再炖）
unfamiliar /ˌʌnfəˈmɪliə(r)/ **adj.** 不熟悉的
combination /ˌkɒmbɪˈneɪʃn/ **n.** 组合
memorize /ˈmeməraɪz/ **v.** 记住
term /tɜːm/ **n.** 术语

中文

David：这些是你需要知道的基本烹饪技法：油炸、翻炒、炙烤、烘烤、烘焙、沸水煮、低温水煮、蒸、煨、炖煮和焖煮。

Steve：我不熟悉这些烹饪术语。

David：我理解。有些烹饪技法使用干热，有些使用湿热，而另一些则结合了两者。

Steve：我明白。

David：有些菜可能会使用多种烹饪技法。但首先，你需要记住这些术语。

Steve：好的，厨师长。

注释

烹饪技法词汇是烹饪英语里的基本内容。先记住以下动词：

Cooking method 烹饪技法			
fry	煎，炒，炸	stir-fry	炒，翻炒
deep-fry	炸（油多）	grill	烤（直接火源）
roast	烤	bake	烘焙
boil	水煮（煮沸）	poach	水煮（水温低一些）
steam	蒸	simmer	煨，慢炖（文火）
stew	炖，焖	braise	焖（先煎再炖）

（1）若不做特别说明，"fry"一词在西餐里一般指油炸这种烹饪的方式，例如"炸鸡"（fried chicken），"炸青瓜"（fried zucchini），"炸鱼"（fried fish）。

1）在翻译"煎，香煎"时，使用"炊具+fried"形式。比如：

pan-fried 平底锅煎　　　　　　　　skillet-fried 长柄煎锅煎

2）在使用不同的油量时。

deep-fried 油炸（油浸过食物）　　shallow-fried 油煎（使用少量烹调油）

3）与动作、速度相关的烹饪技法。

stir-fried 翻炒　　　　　　　　　quick fried 快炒，爆炒

（2）西餐里与"烤"有关的词汇。

roast	烤（烤箱）	roast chicken 烤鸡（整鸡）
bake	烤（烤箱）	baked chicken 烤鸡（鸡胸、鸡腿等）
grill	烤（烤架）	grilled chicken 烧鸡 wood-grilled chicken 柴火鸡
barbeque	烤（户外烤炉）	barbequed chicken 烤鸡（户外烧烤）
broil	炙烤（烈火）	broiled chicken 烤鸡

其中，英语里惯用"roast"来描述烤肉类，烤的过程中会不时地在食材上添加油脂或酱汁；用"bake"来描述烘烤鸡肉、甜点及蔬菜，烹调食物时不添加任何额外的脂肪或油脂。

（3）西餐烹调里，"boil"及"poach"的区别是水温："boil"使用沸腾的水或汤汁来"水煮"食物；"poach"则是指在某一较高温度的液体里进行"水煮"。在翻译中餐菜单时，可根据厨师的烹饪方式来选择词汇。比如：

poached shredded turnip　　　　　水煮萝卜丝
boiled broccoli　　　　　　　　　白灼西蓝花
boiled shrimp　　　　　　　　　　白灼虾

（4）中餐的"炖"一词可以对应英文单词"stew""braise""simmer"；若要细究，西餐里"stew"，指的是用少量的水或汤汁将食物慢慢煮熟，成菜时汤汁与食材一起装碟，"stew"可通译为"炖"；中餐的烹饪技法"焖"可以对应英文词汇"braise"，指先油煎再加大量的水、高汤或酱汁慢炖，在翻译红烧的菜式时选用此词；中餐的烹饪技法"煨"是指用文火慢煮，长时间煨制，可对应英文词汇"simmer"。比如：

stewed beef with potatoes　　　　　土豆炖牛肉
braised beef in red wine　　　　　　红酒炖牛肉
braised beef with brown sauce　　　红烧牛肉
simmered beef with turnips　　　　　萝卜炖牛肉

除了红烧牛肉外，其余三道菜都译为了"食材+炖牛肉"。

（5）菜名翻译的原则要求烹饪技法词汇为过去分词形式，使这些动词有了形容词的词性以描述菜式。比如，"炒肉片"翻译为"fried sliced pork"，这里"fry"（炒）是一个动词，"fried"（炒的）是一个形容词。

2. The Knife Skills 刀工

（初级厨师：Steve，中餐厨师长：David）

David: The size and shape of food materials affect the flavor and texture of a dish.
Steve: I know every shape and size has a different name.
David: Yes. The basic knife cuts are slice, chop, chunk, dice, mince, julienne, etc.
Steve: I want to be skillful with the knife.
David: Many knife cuts are easy to achieve, all you need is to continuously practice. Go ahead!

词汇

knife skill /naif skil/ 刀法，刀工
size /saɪz/ n. 大小，尺寸
shape /ʃeɪp/ n. 形状
material /məˈtɪəriəl/ n. 材料
affect /əˈfekt/ v. 影响
texture /ˈtekstʃə(r)/ n. 口感，质地
knife cut /naɪf kʌt/ 刀工
slice /slaɪs/ n. 片；v. 切片
chop /tʃɒp/ n. 切碎物，剁碎物；v. 切碎，剁碎
chunk /tʃʌŋk/ n. 大块，厚块
dice /daɪs/ n. 丁；v. 切丁
mince /mɪns/ n. 碎末，肉末；v. 切碎
julienne /ˌdʒuːliˈen/ n. 细丝，细条；v. 把（食物切成细丝）
etc. abbr. /ɪtˈsetərə/ 等等 (et cetera)
skillful /ˈskɪlfl/ adj. 熟练的
achieve /əˈtʃiːv/ v. 达到，获得
continuously /kənˈtɪnjuəsli/ adv. 持续不断地
practice /ˈpræktɪs/ v. 练习
go ahead /ˈɡəʊ əhed/ 开始；前进

中文

David：食材的大小和形状会影响菜肴的风味和质地。
Steve：我知道每种形状和尺寸都有不同的名称。
David：是的。基本的刀法切出：片、块、厚块、丁、末、丝等。
Steve：我想要有熟练的刀工。
David：很多刀法不难达到，你需要的是不断地练习。练起来吧！

注释

刀工（knife cut/knife skill）的术语有许多惯用表达，由于在菜单及菜谱里这些术语以动词形式、名词形式（含单复数）和过去式交替使用，需要特别注意：

Knife Cuts 刀工

动词	形容词	名词	名词复数
—	chunky （含有）厚片的，大块的	chunk 厚片，大块，厚块	chunks 大块，厚块
chop 剁碎，砍，切碎	chopped 切碎的，切成块的	chop （羊或猪）排	chops 排骨

（续）

动词	形容词	名词	名词复数
slice 把……切成薄片	sliced （食物）已切成薄片的	slice （切下的食物）薄片	slices 薄片
dice/cube 把（食物）切丁	dicec/cubed 切成丁的	dice/cube 小块食物，（食物）丁	dices(dice)/cubes （食物）丁
mince 切碎，切末	minced 切成末的	mince 末	不可数
julienne 把（食物）切成细丝、细长条	julienne 切成细长条，丝的	julienne 切成细长条、细丝的食物	juliennes 切成细长条、细丝的食物
shred 切丝	shredded 切成丝的	shred 细丝	shreds 细丝
mash 捣碎，捣成糊状（泥）	mashed 捣成湖状（泥）的	mash 糊状物（泥）	mashes 糊状物（泥）

（1）以土豆为例。

菜名：

  土豆块   chunky potatoes/potato chunks
  土豆片   sliced potatoes/potato slices
  土豆丁   diced potatoes/potatoes cubes
  土豆丝   shredded potatoes/julienne potatoes
  土豆泥   mashed potatoes

处理食材：

  将土豆切块   chop the potato
  将土豆切片   slice the potato
  将土豆切丁   dice the potato
  将土豆切成长条 julienne the potato
  将土豆捣成泥 mash the potato

（2）以猪肉为例。

菜名：

  猪肉块   chunky pork/pork chunks
  猪排（带骨） pork chops
  猪肉片   sliced pork/pork slices
  猪肉丁   diced pork/pork cubes
  猪肉丝   shredded pork/pork shreds
  猪肉末   minced pork/pork mince

处理食材：

  将猪肉剁块   cut the pork into chops
  将猪肉切片   slice the pork

将猪肉切丁	dice the pork/cut the pork into cubes
将猪肉切成丝	shred the pork
将猪肉绞碎	mince the pork

在描述"切成……"时需要使用动词的过去分词形式，这些过去分词有了形容词的词性可以描述这些食材的大小及形状。

（3）熟记以下刀工的英文词汇。

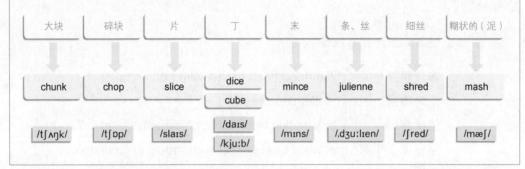

3. The Cookware 炊具

（初级厨师：Steve，中餐厨师长：David）

David: Sometimes we don't transfer food to plates. Pots and pans can just go stove-to-table.
Steve: May I ask, what is "stove-to-table"?
David: Here are some examples: clay pot chicken rice, beef steak served on a sizzling iron plate, stewed crab roe and tofu in stone pot. These dishes are served right on the table without having to transfer the food to a serving platter.
Steve: I see. It's a good idea, we wash fewer dishes.
David: Ha-ha. Know the names of different types of pots and pans and memorize them.
Steve: Yes, chef.

词汇

transfer /trænsˈfɜː(r)/ v. 转移
pots and pans /pɒts ənd pænz/ 坛坛罐罐，锅类总称
stove-to-table /stəʊv tʊ ˈteɪb(ə)l/ 从炉灶到餐桌，带锅上桌
clay pot /kleɪ pɒt/ 砂锅
sizzling iron plate /sɪzlɪŋ ˈaɪən pleɪt/ 铁板（菜）
sizzling /ˈsɪzlɪŋ/ adj. 极热的 v. 发咝咝声（sizzle 的现在分词）
iron /ˈaɪən/ adj. 铁的
crab roe /kræb rəʊ/ 蟹膏，蟹黄
stone pot /stəʊn pɒt/ 石锅
sone /stəʊn/ n. 石头
serving platter /ˈsɜːvɪŋ plætə(r)/ 装菜碟
dishes /ˈdɪʃɪz/ n. 餐具（dish 的复数）
memorize /ˈmeməraɪz/ v. 记住

中文

David：有时候我们不把食物装碟，而让煮锅上餐桌。
Steve：我想问一下，什么是"煮锅上餐桌"？

David: 举些例子：鸡肉煲仔饭、铁板牛排、石钵蟹黄豆腐。这些菜上菜时是带锅上桌，而不必将食物盛到盘子里。

Steve: 我明白了。这是一个好主意，我们可以少洗碗。

David: 哈哈。去了解不同类型的锅碗瓢盆的名称并记住它们。

Steve: 好的，厨师长。

注释

中西餐的菜单上有些菜式的菜名包含有锅具，除了对话里提及的瓦煲、铁板、石锅，还有平底锅、砂锅、汽锅、铁锅等。比如：

蜜汁鸭胸	pan fried duck breast with honey sauce
玉米煎饼	skillet cornbread
脆皮烤鸡	crispy skin oven roast chicken in skillet
锅仔萝卜牛腩	stewed beef brisket with radish in casserole
干锅笋片	griddle cooked bamboo shoots

以下列举了九种菜单上常见的炊具，请尽可能熟记。

中文	英文	音标
平底锅	pan	/pæn/
长柄平底煎锅	skillet	/ˈskɪlɪt/
焙盘、砂锅	casserole	/ˈkæsərəʊl/
砂锅、瓦罐、黏土锅	clay pot	/kleɪ pɒt/
石锅	stone pot	/stəʊn pɒt/
汽锅、蒸锅	steam pot	/stiːm pɒt/
铁锅	griddle	/ˈɡrɪdl/
铁板	sizzling ironplate	/ˈsɪzlɪŋ ˈaɪən pleɪt/
真空低温慢煮（机）	sous vide	/ˌsuːˈviːd/

（1）了解菜式的烹饪方式，根据烹饪方式进行翻译，意译比直译更能让人理解。

（2）理解烹饪方式和食材之间不同翻译版本的区别，进而灵活翻译。例如，酱猪肘的"酱"可以使用"marinated..."，也可以使用"...in brown sauce"；辣椒作为配菜时，可以翻译为"...with chili pepper"或者"...with chili sauce"。

（3）背诵并反复记忆食物词汇最重要的基本功，翻译菜名必须建立在对食品词汇的熟悉与掌握上。

以下是本书针对烹饪英语归纳的菜名翻译原则：

Principles of Menu Translation 菜名翻译原则

1. 烹饪法（动词过去分词）+ 刀工 + 主料 + with/in + 配料

番茄炒牛肉 fried sliced beef with tomato
红酒焖鸭胸 braised duck breast in red wine

2. 主食 + 配菜

鸡腿饭 rice with chicken drumstick
皮蛋瘦肉粥 congee with pork and preserved eggs

3. 主料 with/in + 酱汁

葱油鹅肝 goose liver with scallion oil
蒜汁鹅肝 goose liver in garlic sauce

4. 主菜 +of+ 地名 + style；地名 + 主菜

北京炸酱面 noodles with soy bean paste, Beijing style
广东点心 Cantonese dim sum
北京烤鸭 Beijing roast duck
桂林米粉 Guilin rice noodles

5. 以形状、口感为主，原料为辅

脆皮鸡 crispy chicken
酸甜排骨 sweet and sour pork chops

6. 音译（体现中国餐饮文化）

饺子 jiaozi　　　小笼包 xiao long bao
粽子 zongzi　　　油条 youtiao

7. 意译

龙虎凤大烩 thick soup of snake, wild cat and chicken
霸王别姬 stewed turtle with chicken

8. 典故

东坡肉 dongpo pork
叫花鸡 beggar's chicken

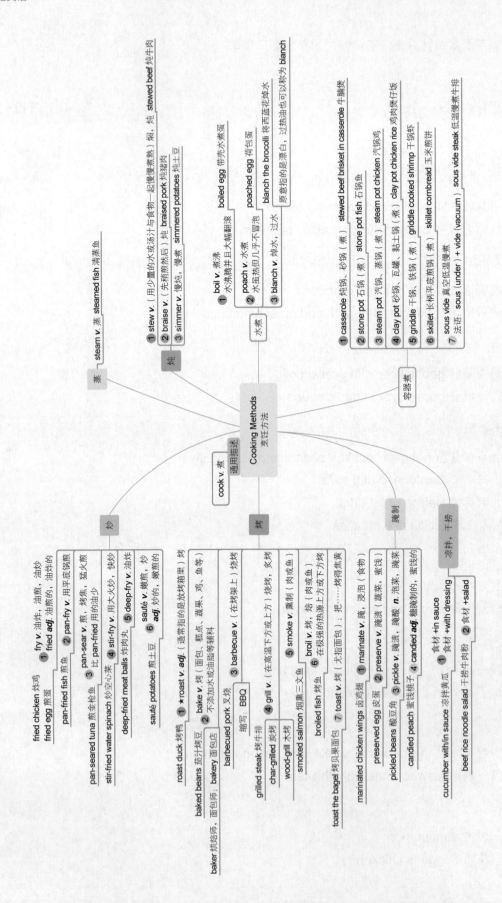

练习：中文菜单翻译

根据前文所述的菜名翻译原则，配合使用手机翻译软件或词典，将以下十道中文菜式翻译为英文。

	Chinese Menu Translation 中文菜单翻译练习	
1	蒜汁鸭胗 烹饪法：凉拌菜 with sauce 材　料：鸭胗 duck gizzard	
2	蒸肉饼 烹饪法：蒸的 steamed 刀　工：切碎的 minced 材　料：猪肉 pork	
3	青椒炒肚丝 烹饪法：炒的 sautéed 刀　工：切片的 sliced 材　料：猪肚 pork tripe，青椒 green pepper	
4	香酥鸭 烹饪法：酥脆 crispy（以口感代替烹饪法） 材　料：鸭肉 duck	
5	姜丝鸭肉 烹饪法：焖的 braised 材　料：鸭肉 duck，姜丝 shredded ginger	
6	蜜汁烤排骨 烹饪法：烤的 roast 材　料：排骨 spare ribs，蜜汁 honey sauce	
7	蒸鲈鱼 烹饪法：清蒸的 steamed 材　料：鲈鱼 perch，姜葱 scallion and ginger	
8	上汤菠菜 烹饪法：先炒后焖的 braised 材　料：菠菜 spinach，上汤 broth	
9	鸡腿饭 主　食：米饭 rice 烹饪法：炖的 stewed 材　料：鸡腿 chicken drumstick	
10	杨枝甘露 烹饪法：冰镇的 chilled 材　料：芒果泥 mango puree，柚子 pomelo，西米露 sago	

职业提示

讲中国故事，展文化实力

燧人氏钻木取火，开启了华夏文明的石烹时代。商朝伊尹提出了药食同源的养生观念。中华美食文化蕴含了阴阳平衡、中庸之道的理念，体现了中国人平稳、和谐的生活心态。随着世界各国的交往日益频繁，同学们应该提高跨文化交际能力，成为立足中华、面向世界的优秀厨师，弘扬中国传统厨艺，与其他国家的美食文化相互借鉴、切磋，开创中华美食文化的新局面。

Unit 12 Meats and Poultry
肉类

肉类词汇是菜单的重要组成部分，除了鸡肉、鸭肉、牛肉、猪肉等，还有不同的部位词汇，比如腿肉、胸肉和内脏等。即使掌握了翻译的方法，若没有足够的词汇量也无法顺利理解英文菜单及菜谱。本单元需要同学们熟记常见的肉类及肉类部位的词汇。

Learning Objectives 学习目标

❊ Memorize the terms of meats and meat part.
  记住肉类及肉类不同部位的词汇。
❊ Practice translating some dishes names.
  练习翻译一些菜名。

扫码看视频

Basic Terms 基础词汇

鸡翅 /ˈtʃɪkɪn wɪŋ/ **chicken wing**

鸡胸 /ˈtʃɪkɪn brest/ **chicken breast**

鸡腿 /ˈtʃɪkɪn ˈdrʌmstɪk/ **chicken drumstick** chicken leg

鸭肉 /dʌk/ **duck**

猪肉 /pɔːk/ **pork**

牛肉 /biːf/ **beef**

胸肉 畜类	胸肉 禽类	肝脏
/ˈbrɪskɪt/ **brisket**	/brest/ **breast**	/ˈlɪvə(r)/ **liver**

肚胃	关节 肘	排骨
/traɪp/ **tripe**	/ˈnʌkl/ **knuckle**	/ˌspeə ˈrɪb/ **spare rib**

腰子 肾脏	胗	里脊肉
/ˈkɪdni/ **kidney**	/ˈɡɪzəd/ **gizzard**	/ˈtendəlɔɪn/ **tenderloin**

肠	猪小肠	脊骨
/ɪnˈtestɪn/ **intestine**	/ˈtʃɪtelɪŋz/ **chitterlings**	/spaɪn/ **spine**

（禽类的）腿	小腿	一大块带骨肉

/θaɪ/ **thigh**	/ʃæŋk/ **shank**	/dʒɔɪnt/ **joint**

内脏 畜类	内脏 禽类	瘦肉

/ˈɒfls/ **offals**	/ˈdʒɪbləts/ **giblets**	/liːn miːt/ **lean meat**

五花肉	肥牛 雪花牛肉	猪蹄

/ˈstriːki pɔːk/ **streaky pork**	/ˈmɑːbld biːf/ **marbled beef** streaky beef	/pɪgs fiːt/ **pig's feet** pig's trotter

肉片	尾巴	整只

/ˈfɪlɪt/ **filet**	/teɪl/ **tail**	/həʊl/ **whole**

Functional Expressions 实用表达

流利地朗读以下菜名，并做到中英互译。

1. 烤鸡肉	roast chicken	
2. 烤火鸡	roasted turkey	
3. 炒鸡肉	fried chicken	
4. 炒鸡杂	fried chicken giblets	
5. 炒鸡胗	fried chicken gizzards	

chicken /ˈtʃɪkɪn/ *n.* 鸡肉　　　　　turkey /ˈtɜːki/ *n.* 火鸡肉　　　　giblet /ˈdʒɪblɪt/ *n.* 禽类的内脏
chicken gizzard /ˈtʃɪkɪn ˈgɪzəd/ 鸡胗　　gizzard /ˈgɪzəd/ *n.* （鸟的）砂囊，胗

6. 烤鸡腿	roasted chicken leg
7. 烤鸭肉	roast duck
8. 卤鸡爪	marinated chicken feet
9. 卤鸡翅	marinated chicken wings
10. 卤鸭脖	braised duck neck

leg /leg/ ***n***. 腿　　　　　　　　　　duck /dʌk/ ***n***. 鸭肉　　　　feet /fi:t/ ***n***. 脚（foot 的复数）
wing /wɪŋ/ ***n***. 翼，翅膀　　　　　　neck /nek/ ***n***. 脖子

11. 焖鸭胸	braised duck breast
12. 炖牛肉	stewed beef
13. 炖牛腩	stewed beef brisket
14. 炒牛百叶	stir-fried beef tripe
15. 焖五花肉	braised pork belly

breast /brest/ ***n***. 胸部　　　　　　　beef /bi:f/ ***n***. 牛肉　　　　　beef brisket 牛腩
brisket /'brɪskɪt/ ***n***. （牛等的）胸脯肉　beef tripe /bi:f traɪp/ 牛百叶，牛的肚
tripe /traɪp/ ***n***. 食用牛肚（或猪肚）；百叶　pork belly /pɔ:k 'beli/ 五花肉，猪腹部的肉
belly /'beli/ ***n***. 腹部；肚子；胃

16. 蒸排骨	steamed pork ribs
17. 炭烧排骨	char-grilled spare ribs
18. 干煸肥肠	stir-fried pork intestines
19. 炒猪肝	fried pork liver
20. 爆炒腰花	quick fried pork kidney

pork /pɔ:k/ ***n***. 猪肉　　　　　　　　pork rib /pɔ:k rɪb/ 猪肋排，排骨
rib /rɪb/ ***n***. 肋骨；排骨　　　　　　 spare rib /speə(r) rɪb/ 排骨，大部分肉都被切去的猪肉排骨
pork intestine /pɔ:k ɪn'testɪn/ 猪肠
intestine /ɪn'testɪn/ ***n***. （脊椎动物的）肠；（无脊椎动物的）肠管
pork liver /pɔ:k 'lɪvə(r)/ 猪肝　　　　 liver /'lɪvə(r)/ ***n***. （动物供食用的）肝
pork kidney /pɔ:k 'kɪdni/ 猪腰，猪肾　　kidney /'kɪdni/ ***n***. （食用的）动物腰子

21. 炖猪骨	stewed pork bones
22. 焖猪肘	braised pork knuckle
23. 炖羊肉	stewed lamb
24. 焖羊小腿	braised lamb shanks
25. 烤羊排	grilled lamb chops
26. 蒸鱼	steamed fish

pork bone /pɔ:k bəʊn/ 猪骨　　　　　　bone /bəʊn/ ***n***. 骨头
pork knuckle /pɔ:k 'nʌkl/ 猪肘　　　　 knuckle /'nʌkl/ ***n***. （猪等动物的）肘，蹄
lamb /læm/ ***n***. 羊羔肉　　　　　　　lamb shank /læm ʃæŋk/ 羊小腿
shank /ʃæŋk/ ***n***. （动物或人的）胫，小腿　lamb chop /læm tʃɒp/ 羊排，羊骨块
chop /tʃɒp/ ***n***. 猪（或羊等）排　　　　fish /fɪʃ/ ***n***. 鱼肉

Task 任务　单词背诵

背诵肉类词汇。

考核：同学们以组为单位，互相抽查词汇记忆。

Dialogues 对话

1. Poultry Meat 禽肉

（初级厨师：Steve，中餐厨师长：David）

David: Steve, go get some eggs from the poultry section from the storage.
Steve: What is poultry?
David: Poultry is the meat from these birds: chickens, ducks, turkeys, geese, pigeons, quails, etc.
Steve: Oh, I see.
David: There is no banquet without chicken. There are many chicken and duck dishes in Chinese cuisine. Geese and pigeons are also common poultry meats.
Steve: I get it, Chef.

词汇

poultry meat /ˈpəʊltri miːt/ 家禽肉
poultry /ˈpəʊltri/ *n.* 家禽；禽肉
meat /miːt/ *n.* 肉，肉类（食用）
egg /eg/ *n.* 鸡蛋，蛋
section /ˈsekʃn/ *n.* 区域
storage /ˈstɔːrɪdʒ/ *n.* 仓库，贮藏室
bird /bɜːd/ *n.* 鸟，禽类
turkey /ˈtɜːki/ *n.* 火鸡，火鸡肉
goose /guːs/ *n.* 鹅，鹅肉
geese /giːs/ *n.* 鹅（goose 的复数形式）
pigeon /ˈpɪdʒɪn/ *n.* 鸽子，鸽肉
quail /kweɪl/ *n.* 鹌鹑，鹌鹑肉
etc. /ɪt ˈsetərə/ 等等
common /ˈkɒmən/ *adj.* 常见的，普通的

中文

David：史蒂夫，去仓库的禽肉区拿些鸡蛋。
Steve：什么是禽肉？
David：禽肉是这些禽类的肉：鸡、鸭、火鸡、鹅、鸽子和鹌鹑等。
Steve：哦，我知道了。
David：无鸡不成宴。中餐里有很多鸡肉和鸭肉菜肴。鹅和鸽子也是常见的家禽肉。
Steve：我懂了，厨师长。

注释

家禽及家畜类肉质可口，营养丰富，适合多种烹调方法。不仅是中餐的重要食材，在西餐的菜单里也占据了大部分篇幅。请记住以下常见与家禽类有关的词汇。

Poultry			
chicken /ˈtʃɪkɪn/	鸡肉	duck /dʌk/	鸭肉
goose /guːs/	鹅肉	turkey /ˈtɜːki/	火鸡
pigeon /ˈpɪdʒɪn/	鸽子，鸽肉	quail /kweɪl/	鹌鹑
spring chicken /ˌsprɪŋ ˈtʃɪkɪn/	稚鸡，童子鸡	silkie (chicken) /ˈsɪlki/	乌鸡
yolk /jəʊk/	蛋黄	egg white /eg waɪt/	蛋白

2. The Livestock 家畜

（初级厨师：Steve，中餐厨师长：David）

David: Livestocks are farm animals.
Steve: What are they?
David: The category includes primarily cattle, sheep, pig, horse, donkey, and mule.
Steve: I see. Is beef a kind of livestock?
David: Yes. Beef is the meat from cattle. The meat from a pig is called pork. Mutton is the meat from an adult sheep, and lamb is the meat from a young sheep. You need to memorize these terms.
Steve: Yes, chef.

词汇

livestock /ˈlaɪvstɒk/ *n*. 家畜，牲畜
farm /fɑːm/ *n*. 养殖场，农场
animal /ˈænɪml/ *n*. 动物
category /ˈkætəɡəri/ *n*. 种类，分类
include /ɪnˈkluːd/ *v*. 包含
primarily /ˈpraɪmərəli/ *adv*. 主要地
cattle /ˈkæt(ə)l/ *n*. 牛的统称
sheep /ʃiːp/ *n*. 羊，绵羊（单复数同形）
pig /pɪɡ/ *n*. 猪
horses /ˈhɔːsɪz/ *n*. 马（horse 的复数形式）
donkey /ˈdɒŋki/ *n*. 驴
mule /mjuːl/ *n*. 骡
mutton /ˈmʌtn/ *n*. 羊肉
adult /ˈædʌlt/ *adj*. 成年的；成熟的
lamb /læm/ *n*. 小羊；羔羊肉
memorize /ˈmeməraɪz/ *v*. 记住
term /tɜːm/ *n*. 术语

中文

David：家畜属于饲养动物。
Steve：都有什么呢？
David：类别主要包括牛、羊、猪、马、驴和骡。
Steve：我明白了。牛肉是畜肉的一种吗？
David：是的。牛肉是牛的肉。猪的肉叫作猪肉。羊肉是成年羊的肉，而羔羊肉是小羊的肉。你需要记住这些术语。
Steve：好的，厨师长。

> **注释**
>
> 在英语里，"meat"是可食用肉类的统称，包括来自猪、牛、羊等动物的肉，但是不包括鱼、海产品及家禽肉。因此，鸡肉和鸭肉等家禽肉"poultry"与家畜肉"meat"分属于两个类别。请记住以下与家畜类有关的词汇。

Meat			
pork /pɔːk/	猪肉	beef /biːf/	牛肉
steak /steɪk/	牛排	veal /viːli/	小牛肉
lamb /læm/	羔羊肉	mutton /mʌtn/	羊肉
venison /venɪsn/	鹿肉	horse meat /hɔːs miːt/	马肉
donkey meat /dɒŋki miːt/	驴肉	rabbit meat /ræbɪt miːt/	兔肉

3. Different Cuts of Meat 肉的不同部位

（初级厨师：Steve，中餐厨师长：David）

David: Different cuts of meat have different names.
Steve: Do you mean "wing", "leg" and something like that?
David: Yes. Take chicken as an example, the primal cuts are neck, wing, breast, tenderloin, thigh, leg, and drumstick.
Steve: What about beef cuts?
David: These are the primal beef cuts: chuck, rib, short rib, brisket, short loin, sirloin, and so on.
Steve: That's quite a lot of terms.
David: Yes, just take your time.

> **词汇**
>
> cuts /kʌts/ *n*. （从动物躯体上）割下的肉块（cut 的复数）
> wing /wɪŋ/ *n*. 翼，翅膀
> leg /leg/ *n*. 腿
> primal /ˈpraɪml/ *adj*. 原始的，主要的
> neck /nek/ *n*. 颈项，脖子
> breast /brest/ *n*. 胸部
> tenderloin /ˈtendəlɔɪn/ *n*. 里脊肉，嫩腰肉
> thigh /θaɪ/ *n*. 鸡腿块，腿股肉；食用的鸡（等的）大腿
> drumstick /ˈdrʌmstɪk/ *n*. 鸡大腿肉；腿下段（像鼓槌的部分）
> chuck /tʃʌk/ *n*. 牛颈肉
> rib /rɪb/ *n*. 肋骨；排骨
> short rib /ʃɔːtrɪb/ 牛小排，牛仔骨
> brisket /ˈbrɪskɪt/ *n*. （牛等的）胸脯肉
> short loin /ʃɔːt lɔɪn/ 里脊肉，前腰肉
> loin /lɔɪn/ *n*. 腰肉
> sirloin /ˈsɜːlɔɪn/ *n*. 后腰肉，西冷

> **中文**
>
> David：不同肉的部位有不同的名称。
> Steve：你是说"翅膀""腿"之类的词吗？
> David：是的。以鸡肉为例，主要的部位有颈部、鸡翅、鸡胸、鸡柳、鸡腿块、全腿和大腿。
> Steve：那牛肉的部位都有哪些呢？
> David：主要的牛肉部位有颈肉、排骨、牛小排、牛腩、里脊肉、外脊肉等。
> Steve：有那么多术语啊。
> David：是的，慢慢来就好。

注释

与肉类词汇比，肉类部位的词汇更繁杂。去除不可食用的部分，如骨头和结缔组织，以下例举常见的不同肉类的切块名称及内脏所对应的英文词汇。

Meat Cuts and Offals			
primal cuts /ˈpraɪml kʌts/	原切块	head /hed/	头部
neck /nek/	颈项，脖子	chuck /tʃʌk/	牛颈肉
wing /wɪŋ/	翼，翅	shoulder /ˈʃəʊldə(r)/	前腿连肩肉（猪、羊）
wing tip /wɪŋ tip/	翅尖	loin /lɔɪn/	腰肉（猪、羊）
breast /brest/	胸脯肉（多用于鸡、鸭）	short loin/sirloin /ʃɔːtlɔɪn/ /ˈsɜːlɔɪn/	前腰 / 后腰肉（牛）
tenderloin /ˈtendəlɔɪn/	里脊肉，嫩腰肉	rib /rɪb/	肋骨，排骨
thigh /θaɪ/	腿股肉，鸡腿块	brisket /ˈbrɪskɪt/	胸脯肉，牛腩（多用于牛、羊、猪）
leg /leg/	全腿，整只腿肉	shank /ʃæŋk/	小腿，胫骨肉（牛、羊）
drumstick /ˈdrʌmstɪk/	大腿肉	hock/knuckle /hɒk/ /ˈnʌkl/	蹄，肘关节
back /beek/	背部，脊柱	round /raʊnd/	臀肉（牛）
offals /ˈɒflz/	内脏（畜类）	giblets /ˈdʒɪbləts/	内脏（禽类）
liver /ˈlɪvə(r)/	肝脏	heart /hɑːt/	心脏
intestine /ɪnˈtestɪn/	肠	chitterlings /ˈtʃɪtəlɪŋz/	猪小肠
tripe /traɪp/	肚，百叶（牛、猪）	belly /ˈbeli/	腹部的肉，五花肉
kidney /ˈkɪdni/	腰子（肾脏）	gizzard /ˈɡɪzəd/	胗，砂囊（禽类）
tongue /tʌŋ/	舌	lung /lʌŋ/	肺
tendon /ˈtendən/	蹄筋，跟腱	tail /teɪl/	尾巴
feet/web /fiːt/ /web/	爪 / 掌蹼	joint /dʒɔɪnt/	一大块带骨肉
boneless /ˈbəʊnləs/	无骨的	spine /spaɪn/	脊骨（龙骨）
carcase /ˈkɑːkəs/	骨架	skin /skɪn/	皮，毛皮
whole /həʊl/	全只，整只	chop /tʃɒp/	肉块，带骨肉块
cullet /ˈkʌtlət/	炸肉排，炸肉片	fillet /ˈfɪlɪt/	无骨肉片（牛、鱼）
lean meat /liːn miːt/	精肉，瘦肉	fat meat /fæt miːt/	肥肉

练习：肉类菜肴翻译

根据前文所提供的肉类及肉类部位词汇，将以下八道肉类菜肴翻译为英文。

	Meat Dishes Translation 肉类菜肴翻译练习	
1	烤羊腿 烹饪法：烤的 roast 肉　类：羊 lamb 部　位：小腿 shank	
2	酱猪蹄 烹饪法：酱焖的 braised...with sauce 肉　类：猪 pork 部　位：肘，关节 hock/knuckle	
3	炸鸡翅 烹饪法：炸的 fried/deep-fried 肉　类：鸡 chicken 部　位：翅膀 wings	
4	蜜汁排骨 烹饪法：焖的 braised 肉　类：猪 pork 部　位：肋骨 ribs/spare ribs	
5	蒸肉饼 烹饪法：蒸的 steamed 刀　工：切碎的 minced 肉　类：猪 pork	
6	红酒焖鸭胸 烹饪法：炖的 stewed/braised 肉　类：鸭 duck 部　位：胸脯肉 breast	
7	鹅掌焖六头鲍 烹饪法：焖的 braised 肉　类：鹅 goose，六头鲍 6-head abalone 部　位：掌 web	
8	花旗参炖乌鸡 烹饪法：隔水炖煮 double boiled（用于翻译中餐的炖盅汤），汤 Soup 或　　：炖的 stewed 肉　类：乌鸡 silkie chicken 材　料：花旗参 ginseng	

职业提示

营养均衡，养生之道

"谷肉果蔬，食养尽之。"中国古人烹饪佳馔，不仅注重美味，还兼顾营养均衡。他们认为每类食物都有各自的特点、功能，按照一定的比例进行搭配，才能利于消化、充分吸收。2016年发布的"中国居民平衡膳食宝塔"更是精细地把谷类、蔬果、鱼禽肉蛋、奶豆制品和调味品进行了分层，标注了建议的摄入量。参照"中国居民平衡膳食宝塔"安排日常饮食，有助于维护人们的健康，减少慢性病的发生。

Unit 13 Aquatic Products
水产品

我国海域广阔，江河纵横，拥有大量的海产品与河产品。中国八大菜系中有不少包含有海、江及河产品的经典菜式，比如松鼠鳜鱼、西湖醋鱼、清蒸石斑鱼、蟹肉烩鱼翅等。全世界海鲜的品种数目巨大，有鱼类、虾类、蟹类、软体类及贝壳类，本单元将对与这些食材相关菜肴的内容进行专门学习。

Learning Objectives 学习目标

❄ Learn the terms of seafood and freshwater products.
学习海产及淡水食物的词汇。

❄ Memorize some common seafood and freshwater product vocabularies.
记住一些常见的海鲜和淡水产品词汇。

❄ Practice translating some dishes names.
练习翻译一些菜名。

扫码看视频

Basic Terms 基础词汇

鲤鱼 /kɑːp/ carp

鳕鱼 /kɒdfɪʃ/ codfish

石斑鱼 /ˈɡruːpə(r)/ grouper

Unit 13 Aquatic Products 水产品 /143

多宝鱼

/ˈtɜːbət/
turbot

秋刀鱼

/ˈsɔːri/
saury

比目鱼

flounder 比目鱼，鲽鱼
/ˈflaʊndə(r)/
halibut 大比目鱼
/ˈhælɪbət/
sole 鳎鱼　　**flatfish** 鲽鱼
/səʊl/　　　　/ˈflætfɪʃ/

甲壳类水生动物

/ˈʃelfɪʃ/
shellfish

鲍鱼

/ˌæbəˈləʊni/
abalone

青口贝
贻贝，淡菜

/ˈmʌsl/
mussel

海螺

/welk/
whelk
sea-snail

扇贝

/ˈskɒləp/
scallop

生蚝

/ˈɔɪstə(r)/
oyster

象拔蚌

/ˈdʒiːəʊ, dʌk klæm/
geoduck clam

蛤蜊

/klæm/
clam

田螺

/ˈrɪvə(r) sneɪl/
river snail

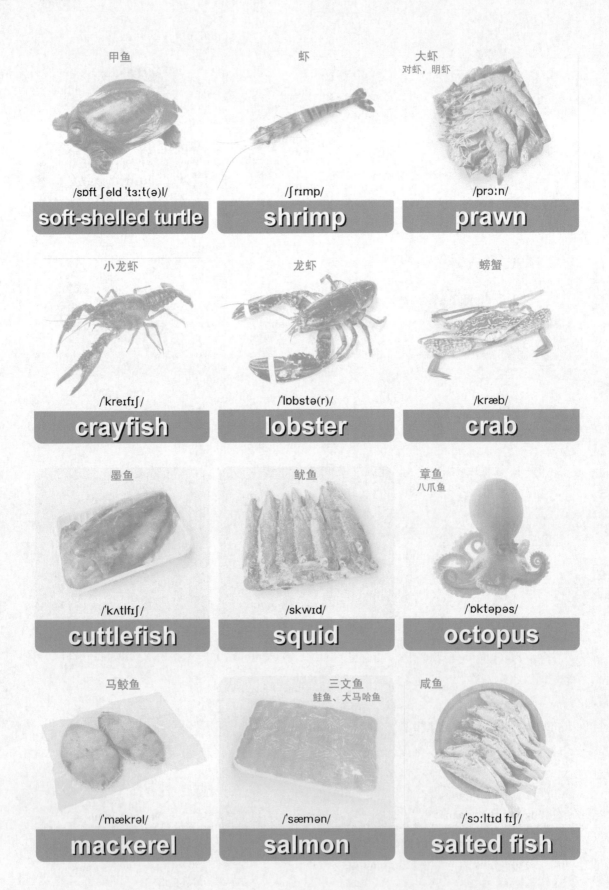

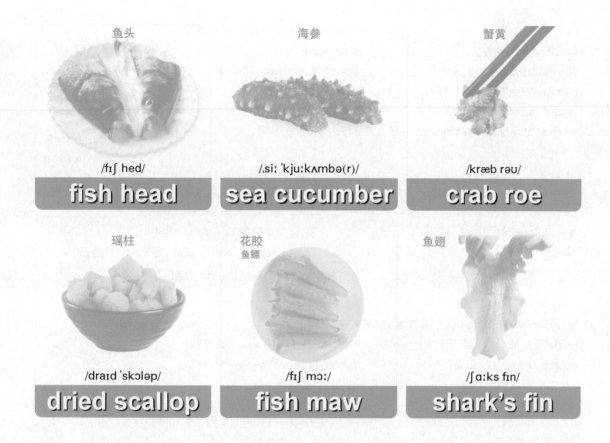

Functional Expressions 实用表达

流利地朗读以下句子，并做到中英互译。

1. 把鱼冲洗一下。	Rinse the fish.
2. 将鱼去除内脏和鳞。	Gut and scale the fish.
3. 鱼已经去鳞了。	The fish is scaled.
4. 用鱼剪去除内脏。	Use fish scissors to gut the fish.
5. 煎之前先用生姜来腌鱼。	Season the fish with ginger before frying it.

rinse /rɪns/ **v**. 冲洗
scale /skeɪl/ **v**. 刮鳞，去鳞
fish scissors /fɪʃ ˈsɪzəz/ 鱼剪
season /ˈsiːzn/ **v**. 给……调味，加作料

gut /gʌt/ **v**. 取出内脏；**n**. 内脏，肠子
scaled /skeɪld/ **adj**. 已去了鳞的
scissors /ˈsɪzəz/ **n**. 剪刀
ginger /ˈdʒɪndʒə(r)/ **n**. 姜

6. 我们可以试试用白葡萄酒来腌鳕鱼。	We can try seasoning the cod with white wine.
7. 你得处理一只活的螃蟹。	You need to deal with a live crab.
8. 把鲜虾放进沸水中。	Place live shrimps in boiling water.
9. 虾变成了粉红色，煮透了。	The shrimp are pink and cooked through.
10. 用龙虾和蟹壳来做海鲜高汤。	Use lobster and crab shells to make seafood stock.

cod /kɒd/ **n**. 鳕鱼
live /laɪv/ **adj**. 活的
shrimp /ʃrɪmp/ **n**. 虾，小虾
pink /pɪŋk/ **adj**. 粉红色的
lobster /ˈlɒbstə(r)/ **n**. 龙虾
seafood stock /ˈsiːfuːd stɒk/ 龙虾高汤

white wine /ˌwaɪt ˈwaɪn/ 白葡萄酒
crab /kræb/ **n**. 蟹
boiling /ˈbɔɪlɪŋ/ **adj**. 沸腾的
cooked through /kʊkt θruː/ 煮熟的
shell /ʃel/ **n**. 壳，贝壳
stock /stɒk/ **n**. 高汤，原汤

11. 蛤和贻贝开口了。 The clams and mussels are opened.
12. 马赛鱼汤是一种海鲜炖菜。 Bouillabaisse is a seafood stew.
13. 海鲜饭是一道经典的西班牙菜。 Paella is a classic Spanish dish.
14. 将虾、蛤、扇贝加入炖鱼里。 Add shrimp, clams, and scallops to the fish stew.
15. 不要把扇贝煮过头了。 Do not overcook the mussel.

clam /klæm/ **n**. 蛤，蛤蜊
bouillabaisse /ˈbuːjəbeɪs/ **n**. 马赛鱼汤，法式杂鱼汤
paella /paɪˈelə/ **n**. （西班牙的）肉菜饭
scallop /ˈskɒləp/ **n**. 扇贝

mussel /ˈmʌsl/ **n**. 贻贝，青口
Spanish /ˈspænɪʃ/ **adj**. 西班牙的
overcook /ˌəʊvəˈkʊk/ **v**. 煮过头

16. 把活鱼养在水里。 Keep the fish alive in the water.
17. 把鱼冰起来。 Freeze the fish.
18. 尽量保持海鲜的味道和口感。 Preserve the taste and texture of seafood as possible as we can.
19. 清蒸鱼上撒有姜丝、葱丝和香菜。 The steamed fish is topped with shredded ginger, scallions, and cilantro.
20. 意大利面配青口贝很好吃。 Spaghetti and mussels are delightful.
21. 龙虾浓汤味道香浓可口。 Lobster bisque is very rich and delicious.

freeze /friːz/ **v**. 冷冻，冻结
taste /teɪst/ **n**. 味道
shredded /ˈʃredɪd/ **adj**. 切碎的
cilantro /sɪˈlæntrəʊ/ **n**. 香菜，芫荽
lobster bisque /ˈlɒbstə(r) bɪsk/ 龙虾浓汤

preserve /prɪˈzɜːv/ **v**. 保存，维持；腌
texture /ˈtekstʃə(r)/ **n**. 口感，质地
scallion /ˈskæliən/ **n**. 葱
spaghetti /spəˈɡeti/ **n**. 意大利细面条
bisque /bɪsk/ **n**. 浓汤（贝壳类）

Task 任务　设计菜单

同学们以组为单位，参考第 11 单元"菜名翻译技巧"及常见水产品词汇，设计一道海鲜大菜，并互相报出自己的菜名。

Dialogues 对话

1. Freshwater Fish 淡水鱼

（初级厨师：Steve，中餐厨师长：David）

David: What's your favorite way to cook fresh water fish?
Steve: Steaming, it's easy. It's a lot more difficult to fry a fish.
David: For pan-fried fish, make sure the fish fillets are well coated with batter and oil.
Steve: I see. Fried fish is delicious.
David: I would say the best way of eating freshwater fish is to barbecue at the side of the river.
Steve: Wow. So freshwater fish is the river fish.
David: Yes. Compare with sea fish, freshwater fish is a bit bony, in general.
Steve: That's why people like sea fish better, in general.

词汇

freshwater /ˈfreʃwɔːtə(r)/ *adj*. 淡水的；*n*. 湖水，内河
favorite /ˈfeɪvərɪt/ *adj*. 最喜欢的
fillet /ˈfɪlɪt/ *n*. 去骨鱼片
coated /ˈkəʊtɪd/ *adj*. 覆有外层或薄膜的
batter /ˈbætə(r)/ *n*. （用鸡蛋、牛奶、面粉等调成的）糊状物
oil /ɔɪl/ *n*. 油
barbeque /ˈbɑːbəkjuː/ *v*. 户外烧烤
side /saɪd/ *n*. 旁边
river /ˈrɪvə(r)/ *n*. 江，河
compare with /kəmˈpeə(r) wɪð/ 与……相比较
bony /ˈbəʊni/ *adj*. 骨瘦如柴的；多骨的
in general /ɪn ˈdʒenrəl/ 通常

中文

David：你最喜欢用什么方式烹饪淡水鱼？
Steve：蒸的，（因为）很简单。煎一条鱼要难得多。
David：煎鱼的话，要确保鱼片上涂有面糊和油脂。
Steve：我明白了。煎鱼很好吃。
David：我认为吃淡水鱼的最好方式是在河边吃烤鱼。
Steve：哇。所以淡水鱼就是河里的鱼。
David：是的。与海水鱼相比，淡水鱼通常刺更多。
Steve：这就是为什么人们通常更喜欢海水鱼的原因。

注释

根据鱼类的生存环境划分，鱼类分为淡水鱼和海水鱼。淡水鱼生活在淡水水域里。我国大部分地区都不临海，但是江河湖泊众多，水系丰富，淡水生物品种繁多。中餐里的鱼、河虾、蟹和蚌等菜式翻译为英文时通常取其品种的大类名，例如：鲤鱼科的草鱼（grass carp）、大头鱼（bighead carp）、鲢鱼（silver carp）、黑皖鱼（black carp）的译名里都有"carp"；把福寿螺、田螺、石螺都翻译为"river snail"；把香螺、花螺、海螺都翻译为"sea whelk"。

在词汇的学习上还要注意，中文和英文对于鲽鱼类鱼种有不同的习惯叫法。鲽鱼也叫比目鱼（flatfish），具有扁平的身体，眼睛只生长在身体的一侧（比目）。中餐菜单里还有鳎鱼、多宝鱼、龙利鱼、鲆鱼、鞋底鱼等词汇，不同地区的人们对比目鱼的称呼还未完全统一，因此本书选择了几个西餐及中餐里常见的比目鱼词汇，在记忆时需要参考这些鱼种的图片以免混淆。

以下是常见的食用鱼所对应的英文词汇。

Fish			
carp /kɑːp/	鲤鱼	silver carp /ˈsɪlvə(r) kɑːp/	鲢鱼
pomfret /ˈpɒmfrɪt/	鲳鱼	crucian carp /ˈkruːʃən kɑːp/	鲫鱼
mandarin fish /ˈmændərɪn fɪʃ/	鳜鱼（桂鱼）	catfish /ˈkætfɪʃ/	鲶鱼
grouper /ˈɡruːpə(r)/	石斑鱼	yellow croaker /ˈjeləʊ ˈkrəʊkə(r)/	黄花鱼
bass /beɪs/	鲈鱼（海）	mackerel /ˈmækrəl/	马鲛鱼
perch /pɜːtʃ/	鲈鱼（河）	tilapia /tɪˈlæpiə/	罗非鱼
herring /ˈherɪŋ/	鲱鱼	hairtail /ˈheəteɪl/	带鱼
sea bream /siː briːm/	海鲷，鲷鱼	snapper /ˈsnæpə(r)/	红鲷，鲷鱼
saury /ˈsɔːri/	秋刀鱼	salmon /ˈsæmən/	三文鱼
tuna /ˈtjuːnə/	金枪鱼，鲔鱼，吞拿鱼	flatfish /ˈflætfɪʃ/	比目鱼，鲽鱼（鲽鱼类统称）
halibut /ˈhælɪbət/	大比目鱼（大）	sole /səʊl/	鳎鱼（扁、小）
turbot /ˈtɜːbət/	多宝鱼，大菱鲆（圆）	flounder /ˈflaʊndə(r)/	鲽，比目鱼，龙利鱼（鳎鱼外的比目鱼统称）
sardine /ˌsɑːˈdiːn/	沙丁鱼	trout /traʊt/	鳟鱼
swellfish /ˈswelfɪʃ/	河豚	eel /iːl/	鳗鱼
finless eel /ˈfɪnlɪs iːl/	黄鳝	salted fish /ˈsɔːltɪd fɪʃ/	咸鱼

2. Crab and Shellfish 蟹及贝壳类动物

（初级厨师：Steve，中餐厨师长：David）

David: Edible fish and shellfish have hundreds of varieties. Can you tell a few names of shellfish species?

Steve: Crabs, lobsters, shrimps, and clams. That's all I know.

词汇

crab /kræb/ *n*. 蟹，蟹肉
shellfish /ˈʃelfɪʃ/ *n*. 水生贝壳类动物（如螃蟹、龙虾、蚌和章鱼）
edible /ˈedəb(ə)l/ *adj*. 可食用的
hundreds of /ˈhʌndrədz ɒv/ 好几百
varieties /vəˈraɪətiz/ *n*. 品种（variety 的复数）
species /ˈspiːʃiːz/ *n*. （生物）物种；种类（单复数同形）
lobster /ˈlɒbstə(r)/ *n*. 龙虾

David: Relevant to human diet, there are oysters, prawns, abalones, mussels as well. Note that some guests have shellfish allergies. Here is the shellfish list, read it through.

Steve: Yes, Chef.

shrimp /ʃrɪmp/ ***n***. 虾（两对爪）	
clam /klæm/ ***n***. 蛤，蛤蜊	
relevant to /ˈreləvənt tu/ 与……有关的	
oyster /ˈɔɪstə(r)/ ***n***. 生蚝，牡蛎	
prawn /prɔːn/ ***n***. 虾（三对爪）	
abalone /ˌæbəˈləʊni/ ***n***. 鲍鱼	
mussel /ˈmʌsl/ ***n***. 贻贝，青口贝	
note that /nəʊt ðæt/ 注意，请注意	
allergies /ˈælədʒiz/ ***n***. 过敏反应（allergy 的复数）	
list /lɪst/ ***n***. 列表，清单	

中文

David：有数百种可食用的鱼和贝类。您能说出几种贝壳类动物的名称吗？
Steve：螃蟹、龙虾、虾，还有蛤。我就知道这么多。
David：与人类饮食有关的，还有牡蛎、大虾、鲍鱼、贻贝。注意，有些客人对贝类过敏。这是贝类清单，请仔细阅读。
Steve：好的，厨师长。

注释

（1）"Edible"一词非常实用，在试图描述一些不知道具体英文名称的食物时可以使用它。比如：

可食用的蘑菇	edible mushroom
可食用的菌类	edible fungus
可食用的蔬菜	edible vegetable
可食用的鱼	edible fish

（2）在翻译"虾"时有两个词汇常常混用："prawn"和"shrimp"。其中具体的区别是：

1）shrimp 有两对爪；prawn 有 3 对爪（它们都有 5 对步足）。

2）shrimp 常被翻译为小虾，而 prawn 则是大虾（相对）。

3）在渔业中，shrimp 指海水、淡水或咸水虾，prawn 专指淡水虾（唐森铭. 英语 shrimp 和 prawn 的区别 [J]. 福建水产，1991，（02））。

4）在日常用语里这两个词都是对虾的统称，没有严格的区分。

（3）过敏症"allergy"是餐饮从业者必须要记住的单词。当客人明确指出自己对某些食物过敏时，无论是服务员还是厨师都需要万分小心。它的形容词为"allergic"。在表达对某些食物过敏时可以这样说：

I am allergic to peanut. 我对花生过敏。
I am allergic to alcohol. 我对酒精过敏。
I have shellfish allergy. 我对贝类有过敏反应。
I have allergic reaction to wheat. 我对小麦过敏。
I eat gluten-free. 我吃无麸质食品。
注：无麸质食品指不含小麦、大麦和黑麦的食物。

（4）贝类动物（shellfish）分为甲壳类贝类（crustacean shellfish）和软体类贝类（molluscan shellfish）。甲壳类贝类有坚硬的外壳，比如虾、蟹、龙虾等，软体类贝类有鲍鱼、螺、贻贝、生蚝、鱿鱼等。下面将这些单词汇录成表，以供系统分类学习。

Shellfish			
crab /kræb/	蟹	hairy crab /ˈheəri kræb/	大闸蟹，毛蟹
shrimp /rɪmp/	虾，小虾	prawn /prɔːn/	对虾，大虾
crustacean /krʌˈsteɪʃ(ə)n/	甲壳纲动物	molluscan /mɒˈlʌskən/	软体动物（无脊椎）
abalone /ˌæbəˈləuni/	鲍鱼	lobster /ˈlɒbstə(r)/	龙虾
oyster /ˈɔɪstə(r)/	生蚝	crayfish/crawdad /ˈkreɪfɪʃ/ /ˈkrɔːdæd/	小龙虾
sea snail /siː sneɪl/	海螺（统称）	whelk /welk/	海螺（大）
clam /klæm/	蚌，蛤	geoduck clam /ˈdʒiːəuˌdʌk klæm/	象拔蚌
scallop /ˈskɒləp/	扇贝	mussel /ˈmʌsl/	贻贝，青口
octopus /ˈɒktəpəs/	章鱼，八爪鱼	squid /skwɪd/	鱿鱼
cuttlefish /ˈkʌtlfɪʃ/	墨鱼，乌贼，花枝	sea urchin /ˈsiːɜːtʃɪn/	海胆
sea cucumber /ˌsiː ˈkjuːkʌmbə(r)/	海参	eel /iːl/	鳗鱼
jelly fish /ˈdʒeli fɪʃ/	海蜇	fish fin /fɪʃ fɪn/	鱼鳍
shark's fin /ʃɑːks fɪn/	鱼翅	fish roe /fɪʃ rəu/	鱼子
fish fillet /fɪʃ ˈfɪlɪt/	鱼肉片（去骨）	fish maw /fɪʃ mɔː/	鱼肚，花胶

3. Octopus, Squid and Cuttlefish 章鱼、鱿鱼和墨鱼

（初级厨师：Steve，行政副厨师长：Jackson）

Jackson: Do you know what the difference between octopus, squid and cuttlefish is？

Steve: They all look the same for me.

Jackson: Actually, their flavors and textures are all different from each other.

Steve: What I know is that calamari is squid.

Jackson: Yes. Calamari is "squid" in Italian. When the meat is served as rings, it is always the squid.

词汇

octopus /ˈɒktəpəs/ n. 章鱼，八爪鱼
squid /skwɪd/ n. 鱿鱼
cuttlefish /ˈkʌtlfɪʃ/ n. 墨鱼，乌贼，花枝
flavor /ˈfleɪvə(r)/ n. 风味，滋味
texture /ˈtekstʃə(r)/ n. 口感，质地
calamari /ˌkæləˈmɑːri/ n. 炸鱿鱼
Italian /ɪˈtæliən/ n. 意大利语
rings /rɪŋz/ n. 环状物（ring 的复数）

Steve: Their meat is all tender and firm after cooking.
Jackson: Yes. However, squid is a bit tougher than octopus, and cuttlefish is in the middle of squid and octopus.
Steve: It's a bit hard to remember. I will try hard.

> tender /ˈtendə(r)/ *adj.* 嫩的；柔软的
> firm /fɜːm/ *adj.* 结实的
> however /haʊˈevə(r)/ *adv.* 但是；然而
> tougher /ˈtʌfər/ *adj.* 更硬的（tough 的比较级）
> in the middle of /ɪn ðə ˈmɪdl əv/ 在……中间

中文

Jackson：你知道章鱼、鱿鱼和墨鱼有什么区别吗？
Steve： 在我看来它们长得都一样。
Jackson：其实，它们的味道和质地都是不一样的。
Steve： 我只知道 calamari（炸鱿鱼）就是鱿鱼。
Jackson：是的。calamari 在意大利语中是鱿鱼的意思。被切成环状的都是鱿鱼。
Steve： 它们的肉在烹调后都又嫩又结实。
Jackson：是的。但是鱿鱼肉比章鱼肉硬一点，墨鱼肉（口感）介于鱿鱼和章鱼之间。
Steve： 有点难记。我会努力的。

注释

国内市场里，通常章鱼最贵，墨鱼次之，鱿鱼最便宜。章鱼的身体又圆又软，有 8 条腿，俗称八爪鱼；相较而言较好辨认。墨鱼和鱿鱼都有 10 条腕足，8 短 2 长。当我们不能从外形分辨鱿鱼和墨鱼时，可以根据它们体内的软骨进行区分：鱿鱼的内贝壳是透明薄片，墨鱼的内贝壳是一块白色的硬骨（海螵蛸）。

章鱼	墨鱼	鱿鱼
octopus /ˈɒktəpəs/	cuttlefish /ˈkʌtlfɪʃ/	squid /skwɪd/

列举几道颇受欢迎的菜式：

grilled octopus	烤八爪鱼
pan-seared octopus with vegetable salad	香煎章鱼配蔬菜沙拉
salt and pepper cuttlefish	椒盐花枝片
stuffed squid	酿鱿鱼
fried calamari with wasabi mayo	炸鱿鱼配芥末蛋黄酱
extra-crunchy calamari	特脆鱿鱼圈
quick fried squid rings	爆炒鱿鱼圈

练习：海鲜菜肴翻译

根据前文所提供的鱼类及海鲜词汇，将以下 8 道菜肴翻译为英文。

	Fish and Seafood Dishes Translation 鱼类和海鲜类翻译练习	
1	上汤焗龙虾 烹饪法：慢炖的 simmered 材　料：龙虾 lobster，高汤 broth	
2	龙井虾仁 烹饪法：先油煎再焖 braised 材　料：虾 shrimp，龙井茶 Longjing tea	
3	粉丝蒸扇贝 烹饪法：蒸的 steamed 材　料：扇贝 scallops，粉丝 vermicelli	
4	西湖醋鱼 烹饪法：水煮的（可不翻译烹饪法） 材　料：西湖鱼（草鱼）West Lake fish，醋汁 vinegar gravy	
5	上海油爆虾 烹 饪 法：翻炒的 stir-fried 地方特色：上海的 Shanghai style 材　　料：虾 Shrimp	
6	糖醋鲤鱼 口　感：酸和甜的 sweet and sour 材　料：鲤鱼 Carp	
7	清蒸鲈鱼 烹饪法：蒸的 steamed 材　料：鲈鱼 perch，葱和姜 Scallion and ginger	
8	蟹黄鱼翅羹 烹饪法：炖的 stewed 　　　　汤 soup（省略烹饪法） 材　料：蟹黄 crab roe，鱼翅 shark's fin	

职业提示

绿色环保，保护家国

"不涸泽而渔，不焚林而猎"是古代先人流传下来的经典名句。然而，在人类文明的发展进程中，随着工业的发展，生态环境遭到了破坏，出现了水土流失、海洋污染、气候变化的现象。二十大报告提出，推进绿色发展，实现人与自然和谐共生。为了建设美丽中国，推动构建人类命运共同体，青年学生有义务从自我做起，节约资源，保护环境，构建人与自然的和谐与共生共荣。

Unit 14 Food Tastes and Textures
食物的味道及口感

在饮食上，中国人追求菜肴的色香味俱全，不吝啬于使用修辞去描述及赞美一道佳肴。对于厨师来说，调味、刀工、火候拿捏、食材把握都是基本功。除了"酸、甜、苦、辣、咸"这些基本的味道描述，能够准确地用语言去描述食物的味道也是一项了不起的技能。

Learning Objectives 学习目标

❋ Know food tastes and textures vocabulary.
学习描述食物味道和口感的词汇。

❋ Learn to describe food better in English.
学会用英语更好地描述食物。

扫码看视频

Basic Terms 基础词汇

味道 taste /teɪst/	口感，质地 texture /ˈtekstʃə(r)/	气味 smell /smel/
滋味 flavor /ˈfleɪvə/	嫩的 tender /ˈtendə(r)/	软的 soft /sɒft/
柔滑的 smooth /smuːð/	多汁的 juicy /ˈdʒuːsi/	开胃的 appetizing /ˈæpɪtaizɪŋ/
酸的 sour /ˈsaʊə(r)/	甜的 sweet /swiːt/	苦的 bitter /ˈbɪtə(r)/
辣的 spicy /ˈspaɪsi/	辣的 hot /hɒt/	麻的 numbing /ˈnʌmɪŋ/
刺鼻的 pungent /ˈpʌndʒənt/	脆的 crispy /ˈkrɪspi/	新鲜的 fresh /freʃ/
香的 aromatic /ˌærəˈmætɪk/	蓬松的 fluffy /ˈflʌfi/	鲜的 umami /uːˈmɑːmi/
有嚼劲的 chewy /ˈtʃuːi/	美味的 tasty /ˈteɪsti/	美味的 delicious /dɪˈlɪʃəs/

Functional Expressions 实用表达

流利地朗读以下句子，并做到中英互译。

1. 吃起来怎么样？	How does it taste?
2. 非常好吃。	It's very tasty/delicious.
3. 味道真棒。	It has an excellent flavor.
4. 好香啊。	What a savory smell.
5. 有一种甜甜的味道。	It has a kind of sweet flavor.

taste /teɪst/ **v.** 吃，尝
delicious /dɪˈlɪʃəs/ **adj.** 美味的，可口的
flavor /ˈfleɪvə(r)/ **n.** 味道，滋味，风味

tasty /ˈteɪsti/ **adj.** 美味的，可口的
excellent /ˈeksələnt/ **adj.** 极好的
savory /ˈseɪvəri/ **adj.** 香的；好吃的；咸味的

6. 爽脆、弹牙、有嚼劲。	It tastes crispy, elastic, and chewy.
7. 口感细腻、柔软。	It has a smooth and soft texture.
8. 这个豆腐吃起来口感很细腻。	The tofu tastes smooth.
9. 这个猪肉太嫩了。	This pork is really tender.
10. 一块好的牛排应该是多汁的。	A good steak should be juicy.

tastes /teɪsts/ **v.** 吃，尝（第三人称单数）
elastic /ɪˈlæstɪk/ **adj.** 有弹性的
smooth /smuːð/ **adj.** 光滑的；细腻的
tender /ˈtendə(r)/ **adj.** 嫩的；柔软的

crispy /ˈkrɪspi/ **adj.** 酥脆的
chewy /ˈtʃuːi/ **adj.** 有嚼劲的；难嚼的
soft /sɒft/ **adj.** 柔软的
juicy /ˈdʒuːsi/ **adj.** 多汁的

11. 很辣很劲爆。	It is so spicy and it hurts so good.
12. 你会觉得嘴巴和舌头变得麻麻的。	You may notice your lips and tongue getting numb.
13. 简直入口即化啊。	It almost melted in my mouth.
14. 有鱼肉的味道。	It tastes like fish.
15. 没有味道。	It's tasteless.

spicy /ˈspaɪsi/ **adj.** 辣的；加了许多香料的
numb /nʌm/ **adj.** 麻木的，失去知觉的
mouth /maʊθ/ **n.** 嘴，口
lip /lɪp/ **n.** 嘴唇
melted /ˈmeltɪd/ **v.** 融化；溶解（melt 的过去式）
tasteless /ˈteɪstləs/ **adj.** 无味的
tongue /tʌŋ/ **n.** 舌头

16. 不合我的口味。	It's not to my taste.
17. 有特殊的气味。	It has a peculiar smell.
18. 菜有点油腻。	It's a bit greasy/oily.
19. 不是很好吃。	It doesn't taste good enough.
20. 这道菜太咸了。	The dish is too salty.

peculiar /pɪˈkjuːliə(r)/ **adj.** 奇怪的，怪异的
oily /ˈɔɪli/ **adj.** 含油多的

greasy /ˈgriːsi/ **adj.** 油腻的，多油的
salty /ˈsɔːlti/ **adj.** 咸的

21. 牛奶有一点发酸。　　　　　　The milk is slightly sour.
22. 牛肉用盐和胡椒粉腌过了。　　The beef is seasoned with salt and pepper.
23. 我想吃点清淡的/重口味的。　 I'd like something light/heavy.
24. 再来一碗香喷喷的汤。　　　　And a bowl of aromatic broth, please.

slightly /ˈslaɪtli/ **adv**. 轻微地
season /ˈsiːzn/ **v**. 给……调味，腌渍
pepper /ˈpepə(r)/ **n**. 胡椒；辣椒
aromatic /ˌærəˈmætɪk/ **adj**. 芳香的
sour /ˈsaʊə(r)/ **adj**. 酸的；酸臭的
salt /sɔːlt/ **n**. 盐
light /laɪt/ **adj**. 清淡的　　heavy /ˈhevi/ **adj**. 浓重的
broth /brɒθ/ **n**. 肉汤，高汤

Task 任务　　翻译练习

练习用英语描述某一种食材的味道：

1. 有点（a bit, slightly）　　　味道：酸、甜、苦、辣、咸的
2. 相当（quite, rather, pretty）　口感：滑、脆、嫩、软、入口即化、老（硬）的
3. 非常（very, awfully, really）　感官：好吃、难吃、油腻、浓郁、清淡、变质的
　　　　　　　　　　　　　　　　熟度：生的、没煮熟、煮过头、焦的

食材：苦瓜、薯片、面条、苹果、东坡肉、羊排等。

Dialogues 对话

1. The Steak is Cooked Perfectly 牛排煮得非常完美

（初级厨师：Steve，西餐厨师长：Neil）

Neil: The fillet cut of steak is the most tender cut. It's also the most expensive steak cut. Try it. Take a bite.
Steve: OK. Wow.
Neil: How does it taste?
Steve: It's good. The steak is cooked perfectly.
Neil: It's juicy, tender, and flavorful. Isn't it?
Steve: Yes, it is. It's very yummy!
Neil: Uh-huh. Timing is key. Try the asparagus on the side.
Steve: It's crunchy, and it goes well with cracked black pepper. I've never tasted any other steak as delicious as this.
Neil: Haha.

词汇

fillet cut /ˈfuɪl kʌt/ 菲力牛肉切块
most tender /məʊst ˈtendə(r)/ 最嫩的
expensive /ɪkˈspensɪv/ **adj**. 昂贵的
bite /baɪt/ **n**. 咬；一口 **v**. 咬
perfectly /ˈpɜːfɪktli/ **adv**. 完美地
juicy /ˈdʒuːsi/ **adj**. 多汁的
tender /ˈtendə(r)/ **adj**. 嫩的；柔软的
flavorful /ˈfleɪvəful/ **adj**. 可口的；充满……味道的
yummy /ˈjʌmi/ **adj**. 很好吃的
timing/ˈtaɪmɪŋ/ **n**. 时间的选择
key /kiː/ **n**. 关键
asparagus /əˈspærəgəs/ **n**. 芦笋
crunchy /ˈkrʌntʃi/ **adj**. 脆的，爽脆的
cracked /krækt/ **adj**. 粗磨的；压碎的
black pepper /blæk ˈpepə(r)/ 黑胡椒
tasted /ˈteɪstɪd/ **v**. 吃（或喝）；体验（taste 的过去分词）
as…as…/əz əz/ 像……一样

中文

Neil: 菲力牛排是最嫩的，也是最贵的牛排切块。试一试。咬一口。
Steve: 好吧。哇。
Neil: 味道怎么样？
Steve: 很好。牛排做得很好。
Neil: 多汁、鲜嫩、可口。不是吗？
Steve: 是的。它很好吃！
Neil: 嗯。时机是关键。试试旁边的芦笋。
Steve: 很脆，和碎黑胡椒很配。我从来没有吃过这么好吃的牛排。
Neil: 哈哈。

注释

英文里有许多词可以表示"好吃的、美味的、可口的"，最常见的表达是"delicious"及"tasty"。以下列举一些形容饭菜美味可口的其他词汇：

很好吃的 yummy	/ˈjʌmi/	yummy cake	美味的蛋糕
香甜的 delectable	/dɪˈlektəbl/	delectable raspberries	甘美的覆盆子
甜美多汁的 luscious	/ˈlʌʃəs/	luscious fruit	清甜饱满的水果
咸香美味的 savory	/ˈseɪvəri/	savory zongzi	咸粽子
愉快满意的 delightful	/dɪˈlaɪtfl/	delightful meal	满意可口的饭菜
美味绝妙的 scrumptious	/ˈskrʌmpʃəs/	scrumptious lunch	丰盛美味的午餐

2. Serving Dishes 席间服务

（客人：Bella，资深餐厅服务员：Eric）

Eric: Excuse me, madam. Is everything OK with your meal?
Bella: Oh. It's alright. I like the soup, it is creamy and rich. This mashed potatoes is quite fluffy.
Eric: That's good. How about the burger?
Bella: The burger was left on the barbecue too long and now it's tough. The chicken is really chewy.
Eric: I'm really sorry about that. May I take the plate back to kitchen and bring a newly-made one for you?
Bella: Never mind, it's OK. I'm stuffed.
Eric: How about a cheese platter, with our compliment.
Bella: OK. Thank you.

词汇

creamy /ˈkriːmi/ *adj.* 含乳脂的
rich /rɪtʃ/ *adj.* 浓郁的，油腻的
mashed potatoes /mæʃt pəˈteɪtəʊz/ 土豆泥
fluffy /ˈflʌfi/ *adj.* 蓬松的；松软的
burger /ˈbɜːgə(r)/ *n.* 汉堡包
barbecue /ˈbɑːbɪkjuː/ *n.* 烤架
tough /tʌf/ *adj.* 硬的，肉嚼不烂的
chewy /ˈtʃuːi/ *adj.* 有嚼劲的；难嚼的
plate /pleɪt/ *n.* 碟子
kitchen /ˈkɪtʃɪn/ *n.* 厨房
newly-made /ˈnjuːli meɪd/ 新做的
never mind /ˈnevə(r) maɪnd/ *n.* 别介意，无所谓
stuffed /stʌft/ *adj.* 已经吃饱了的；塞满了的
cheese platter /tʃiːz plætə(r)/ 奶酪拼盘
compliment /ˈkɒmplɪmənt/ *n.* 致意，问候，恭维

中文

Eric: 打扰了，女士。您对用餐还满意吗？
Bella: 噢，还不错。我喜欢这道汤，奶油丰富，味道浓郁。这个土豆泥很松软。
Eric: 那就好。汉堡包怎么样？
Bella: 汉堡包在烤架上烤得太久了，现在很硬。鸡肉好难嚼。
Eric: 我真的很抱歉。我把碟子拿回厨房，再给你拿一份新做的，好吗？
Bella: 不要介意，没关系。我吃饱了。
Eric: 来一份奶酪拼盘怎么样，算是本店小小敬意。
Bella: 好的。谢谢你。

注释

（1）"我吃饱了"的表达有"I am stuffed"（意指非常饱，肚子已没有空间），此外还有以下几种表达：

I'm full.	我饱了。
I couldn't eat another thing.	我真的吃不下了。
There's no room for any more.	没地方塞食物了。
It really sticks to my rib.	我超饱（胃都要戳到肋骨了）。

（2）"With compliment"和"complimentary"都有"免费赠送"的意思，不仅仅是简单的"免费"(free of charge)，更有一种问候及赞美之意。其中作为形容词的"complimentary"更常见些，比如：

complimentary ticket 赠票	complimentary voucher 代金券
complimentary drink 赠饮	complimentary dessert 赠送的甜点

（3）"重新做一份、重新煮"的表达。

We will cook another one for you.	我们会重新给您做一盘。
The chef will re-cook the dish for you.	厨师会重新为您做这道菜。
I will bring a newly-made dish to you.	我去拿一道新做好的菜给您。
I will have the dish changed right away.	我马上换一份给您。
The chef will prepare another dish for you.	厨师会为您准备另一道菜。

（4）酸（sour）、甜（sweet）、苦（bitter）、咸（salty）、鲜（umami）是五种基本味道（味觉）；脆（crisp）、嫩（tender）等是食物在口腔内的感受（触觉）；香（aromatic）是嗅觉；而好吃（tasty）、开胃（appetizing）则是属于感官知觉。以下汇集了食物味道及一些常见的口感中英对照词汇。

Tastes and Textures			
sour /saʊə(r)/	酸的	sweet /swiːt/	甜的
bitter /ˈbɪtə(r)/	苦的	spicy/hot /ˈspaɪsi/ / /hɒt/	辣的
salty /ˈsɑːlti/	咸的	umami /uːˈmɑːmi/	鲜味的
fresh /freʃ/	新鲜的	numbing hot /ˈnʌmɪŋ hɒt/	麻辣的
pungent /ˈpʌndʒənt/	辛辣的，刺鼻的	light/mild /laɪt/ / /maɪld/	清淡的
heavy/strong /ˈhevi/ / /strɒg/	浓重的	rich /rɪtʃ/	浓郁的
creamy /ˈkriːmi/	奶味的，含乳脂的	fruity /ˈfruːti/	果味的
juicy /ˈdʒuːsi/	多汁的	aromatic /ˌærəˈmætɪk/	香的（食物及植物）
scented /ˈsentɪd/	香的（植物）	fragrant /ˈfreɪgrənt/	香的（芬芳的、味道浓烈的）
soft /sɒft/	软的	tough/hard /tʌf/ / /hɑːd/	硬的
tender /ˈtendə(r)/	嫩的	smooth /smuːð/	柔滑的
crisp/crispy /krɪsp/ / /ˈkrɪspi/	脆的	crunchy /ˈkrʌntʃi/	松脆的（吃的时候发出嘎吱嘎吱声响的食物）
melted /ˈmeltɪd/	融化的	fluffy /ˈflʌfi/	蓬松的、毛茸的
elastic /ɪˈlæstɪk/	有弹性的	al dente /ˌælˈdenteɪ/	筋道的（面食、意大利面）
chewy /ˈtʃuːi/	难嚼的，有嚼劲的	appetizing /ˈæpɪtaɪzɪŋ/	开胃的
greasy/oily /ˈgriːsi/ / /ˈɔɪli/	油腻的	tasteless/blend /ˈteɪstləs/ / /blænd/	无味的
burnt /bɜːnt/	烧焦的	spoiled/off/bad /spɔɪld/ / /ɒf/ / /bæd/	过期的
moldy /ˈməʊldi/	发霉的	disgusting /dɪsˈgʌstɪŋ/	恶心的

3. Dietary Needs 饮食需求

（初级厨师：Steve，西餐厨师长：Neil）

Neil: Here comes an order: Linguine with spicy calamari and garlic. It says that the guest doesn't want any pepper or chili in it.

Steve: What to do?

Neil: Sometimes we will modify our recipe to meet the needs of different guests. Either change the ingredients or the cooking technique. We need to make some changes by substituting the chili sauce.

Steve: How about ragu sauce?

Neil: OK, we are on the same page. Let's deal with the calamari first.

Steve: Yes, chef.

词汇

dietary /'daɪətəri/ *adj.* 饮食上的
needs /niːdz/ *n.* 要求，需求（need 的复数）
linguine /lɪŋ'gwiːnɪ/ *n.* 一种意大利扁面条
spicy /'spaɪsi/ *adj.* 辛辣的
calamari /ˌkæləˈmɑːri/ *n.* 鱿鱼（意）
garlic /'ɡɑːlɪk/ *n.* 大蒜
pepper /'pepə(r)/ *n.* 辣椒，胡椒
chili /'tʃɪli/ *n.* 辣椒
modify /'mɒdɪfaɪ/ *v.* 修改；调整
recipe /'resəpi/ *n.* 食谱
either...or... /'aɪðə(r) ɔː(r)/ 要么……要么……
ingredient /ɪn'ɡriːdiənt/ *n.* 原料
technique /tekˈniːk/ *n.* 技巧，技术
substituting /'sʌbstɪtjuːtɪŋ/ *v.* 代替，取代
chili sauce /'tʃɪli sɔːs/ 辣椒酱
ragu /ræˈɡuː/ *n.* 番茄肉酱（意）
on the same page /ɒn ðə seɪm peɪdʒ/ 达成共识；想到一块去了

中文

Neil: 新的单来了：蒜香辣鱿鱼意面。上面显示客人要求不放辣椒。

Steve: 该怎么做？

Neil: 有时我们会根据不同客人的需要修改食谱。要么改变食材，要么改变烹饪法。我们要换掉辣椒酱。

Steve: 用番茄肉酱怎么样？

Neil: 好的，我们想到一块去了。先来处理鱿鱼吧。

Steve: 好的，厨师长。

注释

（1）"chili"和"pepper"都有辣椒之意，其中"chili"一定是辣的辣椒；而"pepper"有"bell pepper"（柿子椒）和"cherry pepper"（甜椒）等不辣的辣椒，还可以指胡椒和胡椒粉，视上下文而译。

salt and pepper　盐和胡椒
steak with black pepper sauce　黑椒牛排（黑胡椒）
fried pork with hot chili pepper　辣椒炒肉（辣的辣椒）

（2）可以搭配味道描述的程度副词。

1）有点　a bit, slightly

　　The beef is a bit salty, but it's delicious.　牛肉有点咸，但很好吃。

　　The drumstick is slightly tough.　鸡腿有些硬（老）。

　　The sauce is a bit sour.　酱汁有点发酸。

2）相当　quite, rather, pretty

　　This grilled tilapia is rather crispy.　这道烤罗非鱼相当酥脆哦。

　　The steak is quite chewy, it must have been cooked for too long.

　　牛排挺难嚼的，一定是煮太久了。

　　This mashed potatoes is quite fluffy.　这个土豆泥很绵软。

3）非常　very, awfully, really

　　The fish is so sour and sweet, very delightful.　这道鱼又酸又甜，很好吃。

　　This curry is very spicy!　这个咖喱非常辣。

　　This is awfully tasty!　这也太好吃了吧！

练习：描述菜肴味道

根据提示词汇试着将句子翻译为英文。

1. 它吃起来怎么样？ 吃，品尝 taste

2. 这块肉相当嫩。 相当 quite；嫩 tender

3. 闻起来很香。 闻 smell

4. 它几乎入口即化。 融化的 melted

5. 我想吃点清淡的。 清淡的 light

6. 这道菜太咸了。 菜 dish；咸的 salty

7. 这个鸡肉有种甜甜的味道。 有一种 have a kind of；味道 flavor；甜的 sweet

8. 这没有味道。 无味的 tasteless

9. 好香啊！ 气味 smell

10. 面条很筋道。 有嚼劲的（面）al dente

职业提示

适时而食，不时不食

饮食讲究季节时令的变化，因季而食，适时而食，是中华饮食文化的优良传统，对于现代中国人的日常饮食生活具有重要的指导意义。根据"春生、夏长、秋收、冬藏"的规律，春天可以多吃生发、辛散之物，忌食酸收之品，利于肝气畅通。夏季炎热，可以食用苦瓜、乌梅汤等去暑食品。长夏之际，应该健脾化湿，适合瓜豆类食材。秋季为了滋阴润燥可以使用莲藕、山药等根茎类食材。冬天适当进食菠菜与冬笋对人体大有裨益。

季节饮食养生文化是中国饮食文化与养生文化的精粹，在"天人合一"整体哲学观指导下，中国古人很早就认识到了人体与自然环境间的密切关系，中国古代哲学、医学都把人的生存与健康放在生态环境中去认识。"应季而食"指导着中国人餐桌的四季变幻，更成为当前餐饮业产品创新的动力源泉。餐饮从业人员可从餐饮产品的创新方面着手，将四季饮食养生文化资源转化为行业的创新发展动力，为倡导健康科学的饮食理念和中国餐饮业的发展做出贡献。

Part 5
Recipe
食 谱

Unit 15　Condiments 调味品 / 164

Unit 16　Common Measure Words in the Recipe 食谱中常见的量词 / 177

Unit 17　Common Functional Expressions in the Kitchen 厨房常用的功能句 / 186

Unit 15　Condiments
调味品

不同的佐料搭配不同的烹饪手法能带来不同的风味，中西式菜肴的风格差异还体现在使用不同的调味品。读懂英文菜谱，从看懂食谱里的"材料"（ingredients）开始。

Learning Objectives 学习目标

❋ Learn basic words and sentences of condiments and seasonings.
学习调味品和调味料的基本单词和短语。

❋ Learn to read the Ingredients of a recipe.
学会阅读食谱中的配料。

扫码看视频

Basic Terms 基础词汇

盐	花生油	酱油
/sɔːlt/	/ˈpiːnʌt ɔɪl/	/ˈsɔɪ sɔːs/
salt	**peanut oil**	**soy sauce**

醋	料酒	糖
/ˈvɪnɪɡə(r)/	/ˈkʊkɪŋ waɪn/	/ˈʃʊɡə(r)/
vinegar	**cooking wine**	**sugar**

Unit 15　Condiments 调味品 /165

蚝油

/ˈɔɪstə ˌsɔːs/
oyster sauce

米酒
/rɪs waɪn/
rice wine
liquor /ˈlɪkə(r)/

玉米淀粉

/kɔːrn stɑːrtʃ/
corn starch

味精

/ˌmɒnəʊˈsəʊdiəm ˈɡluːtəmeɪt/
**monosodium glutamate
MSG**

鸡精
/ˈtʃɪkɪn ˈesns/
chicken essence

葡萄酒
/waɪn/
wine

油醋汁

/ˌvɪnɪˈɡret/
vinaigrette

橄榄油
/ˌɒlɪv ˈɔɪl/
olive oil

辣椒酱

/ˈtʃɪli sɔːs/
chili sauce

蜂蜜

/ˈhʌni/
honey

奶油

/kriːm/
cream

酵母

/jiːst/
yeast

芥末	糖浆	草本植物
		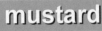
/ˈmʌstəd/	/ˈsɪrəp/	/hɜːb/
mustard	**syrup**	**herb**

香料	黑胡椒	八角
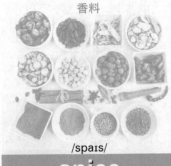		
/spaɪs/	/ˌblæk ˈpepə(r)/	/ˌstɑːr ˈænɪs/
spice	**black pepper**	**star anise**

孜然	肉桂皮	花椒

/ˈkjuːmɪn/	/ˈsɪnəmən/	/ˈsiːtʃwɑːn ˈpepə(r)/
cumin	**cinnamon**	**Sichuan pepper**

肉豆蔻	姜黄	丁香

/ˈnʌtmeg/	/ˈtɜːmərɪk/	/kləʊv/
nutmeg	**turmeric**	**clove**

Unit 15　Condiments 调味品 /167

辣椒	红辣椒粉	卡宴辣椒粉

/ˈtʃɪli ˈpepə(r)/	/ˈpæprɪkə/	/keɪˈen ˈpepə(r)/
chili pepper	**paprika**	**cayenne pepper**

罗勒	藏红花	薄荷

/ˈbæzl/	/ˈsæfrən/	/mɪnt/
basil	**saffron**	**mint** peppermint /ˈpepəmɪnt/

香草	香草精	牛至叶

/vəˈnɪlə/	/vəˈnɪlə ˈekstrækt/	/ˌɒrɪˈɡɑːnəʊ/
vanilla	**Vanilla extract**	**oregano**

香茅草	迷迭香	百里香

/ˈlemən ɡrɑːs/	/ˈrəʊzməri/	/taɪm/
lemon grass	**rosemary**	**thyme**

Functional Expressions 实用表达

请流利地朗读以下句子,并做到中英互译。

1. 准备好所有调味品。 Prepare all condiments.
2. 提前准备好肉类。 Prepare the meat ahead of time.
3. 需要什么食材? What ingredients are needed?
4. 胡萝卜碎、柿子椒丁和土豆泥。 Chopped carrots, diced bell peppers and mashed potatoes.
5. 混合所有的原料。 Combine all ingredients.

prepare /prɪˈpeə(r)/ **v.** 准备
ahead of time /əˈhed ɒv taɪm/ 提前
chopped /tʃɒpt/ **v.** 砍；斩碎（chop 的过去分词）
mashed /smæʃt/ **adj.** 被捣成糊状的

condiment /ˈkɒndɪmənt/ **n.** 调味品
ingredient /ɪnˈɡriːdiənt/ **n.** 原料
diced /daɪst/ **adj.** 切粒的
combine /kəmˈbaɪn/ **v.** 使结合

6. 添加额外的原料。
7. 把肉在冰箱里放一晚上解冻。
8. 把肉放在温水里一小时来解冻。
9. 放入酱油和醋。
10. 给肉调味的时候用粗盐。

Add additional ingredients.
Defrost the meat overnight in the refrigerator.
Defrost the meat with warm water for 1 hour.
Put in soy sauce and vinegar.
Use coarse salt when seasoning meat.

add /æd/ **v.** 加，增加
defrost /diːˈfrɒst/ **v.** 解冻
refrigerator /rɪˈfrɪdʒəreɪtə(r)/ **n.** 冰箱
soy sauce /ˌsɔɪ ˈsɔːs/ 酱油
coarse salt /kɔːs sɔːlt/ 粗盐

additional /əˈdɪʃən(ə)l/ **adj.** 额外的
overnight /ˌəʊvəˈnaɪt/ **adv.** 在夜间
warm water /wɔːm ˈwɔːtə(r)/ 温水
vinegar /ˈvɪnɪɡə(r)/ **n.** 醋

11. 这种调料包含了大蒜和黑胡椒。
12. 在牛肉上撒上盐和黑胡椒。
13. 加多点醋。
14. 在上面撒上罗勒叶。
15. 用鸡汤煮米饭。

This seasoning includes garlic and black pepper.
Sprinkle beef with salt and black pepper.
Add more vinegar.
Sprinkle basil leaves on top.
Cook rice in chicken broth.

seasoning /ˈsiːzənɪŋ/ **n.** 调味品
black pepper /blæk ˈpepə(r)/ 黑胡椒
basil leaves /ˈbæzl liːvz/ 罗勒叶

season /ˈsiːzn/ **v.** 给……调味
sprinkle /ˈsprɪŋk(ə)l/ **v.** 撒；洒
chicken broth /ˈtʃɪkɪn brɒθ/ 鸡汤

16. 品尝汤后再调整味道。
17. 中国人离不开酱油。
18. 生抽是中式烹饪中常用的调料。
19. 蚝油能为菜肴增添鲜味。
20. 玉米淀粉可用来勾芡酱汁及腌肉。

Taste the soup and correct the seasoning.
Soy sauce is a must in the lives of Chinese people.
Light soy sauce is commonly used in Chinese cooking.
Oyster sauce can provide umami flavor to dishes.
Cornstarch can be used for thickening sauces and marinating meat.

correct /kəˈrekt/ **v.** 纠正；改正
oyster sauce /ˈɔɪstə(r) sɔːs/ 蚝油
cornstarch /ˈkɔːnstɑːtʃ/ **n.** 玉米淀粉
thickening /ˈθɪk(ə)nɪŋ/ **v.**（使）变厚，变浓稠（thicken 的现在分词）；**n.** 芡粉
marinating /ˈmærɪneɪtɪŋ/ **v.** 用调料腌渍……；（食物）经腌渍（marinate 的现在分词）

light soy sauce /laɪt sɔɪ sɔːs/ 生抽
umami /uːˈmɑːmi/ **n.** 鲜味

Task 任务　词汇举例

1. List 5 common condiments in Chinese cooking;
   列出 5 种常见的中式烹饪调料
2. List 5 common condiments in Western cooking.
   列出 5 种常见的西式烹饪调料。
3. What condiments are needed to make a braised pork in soy sauce?
   做一道红烧肉需要什么调味品？
4. What condiments are needed to make a grilled steak?
   做一道烤牛排需要什么调味品？

 Dialogues 对话

1. Condiments 调味品

(初级厨师：Steve，中餐厨师长：David)

David: Condiments are essential to any kitchen, they enhance the flavor of dishes.

Steve: A good cook should know when and how to use condiments properly.

David: Well said! Let me ask you, what are the most essential condiments in our kitchen?

Steve: Light soy sauce, oyster sauce, cooking wine, salt, vinegar, and oil, I suppose.

David: Yes. these condiments are each essential to Chinese cooking. Check the pantry and tell me more.

Steve: Yes, chef.

词汇

condiment /ˈkɒndɪmənt/ n. 调味品
essential /ɪˈsenʃl/ adj. 必要的
enhance /ɪnˈhɑːns/ v. 提高；加强
properly /ˈprɒpəli/ adv. 适当地；正确地
light soy sauce /laɪt sɔɪ sɔːs/ 生抽
oyster sauce /ˈɔɪstə(r) sɔːs/ 蚝油
cooking wine /ˈkʊkɪŋ waɪn/ 料酒
salt /sɔːlt/ n. 盐
vinegar /ˈvɪnɪɡə(r)/ n. 醋
oil /ɔɪl/ n. 油
suppose /səˈpəʊz/ v. 认为；推断
pantry /ˈpæntri/ n. 食品室，食品储藏室

中文

David: 调味品对任何厨房来说都是必不可少的，它们能提升菜肴的味道。

Steve: 一个好的厨师应该知道何时以及如何正确地使用调味品。

David: 说得好！我来问问你，我们厨房里最重要的调料是什么？

Steve: 生抽、蚝油、料酒、盐、醋和油，我认为。

David: 是的。这些调料都是中式烹饪中必不可少的。去食品贮藏室看看，然后再告诉我还有什么。

Steve: 好的，厨师长。

注释

"condiment""seasoning""flavoring"都翻译为"调味品、佐料",其中的区别在于:

condiment:指在食物制作的过程中或煮好以后添加的酱汁、香料、盐、油及葱蒜等食材,用来增强食物的味道或添加风味。

seasoning:指食物在煮熟之前就添加进去的食材,比如腌渍肉类的香料、酱汁和葱蒜等蔬菜。动词形式"season"可理解为中餐里所说的"腌,腌制"。

flavoring:指可以添加到食物及饮料里增加风味的材料,比如香草精 (vanilla extract) 和柠檬汁 (lemon juice)。

例句:

Here are the condiments for rice noodles, please help yourself. 这是米粉的调味品,请随意取用。

Add peach as a flavoring to tea. 在茶中加入桃子调味。

Scallion is a widely used seasoning in China. 葱在中国是一种广泛使用的调味品。

2. Herbs ang Spices 香料及香料

(初级厨师:Steve,西餐厨师长:Neil)

词汇

Neil: There are various kinds of herbs and spices used in cooking, the most often used spices are…please give me a few examples, Steve.

Steve: Basil, black pepper, thyme, rosemary, mint, nutmeg, etc. Am I right?

Neil: Yes, you are right. Some of the ones are also frequently used in Chinese cooking, like cinnamon, bay leaf, clove, fennel seed, cumin, chili powder, and so on.

Steve: And some of the ones are classic Chinese spices, like star anise, 5-spice powder, and Sichuan pepper.

Neil: Correct.

various /ˈveərɪəs/ *adj.* 各种各样的
herbs /hɜːbs/ *n.* (调味或药用的) 香草
spices /sˈpaɪsɪz/ *n.* 香味料,调味料(spice 的复数)
most often used /məʊst ˈɔːf(ə)n juːzd/ 最常用的
basil /ˈbæzl/ *n.* 罗勒
black pepper /ˌblæk ˈpepə(r)/ 黑胡椒
thyme /taɪm/ *n.* 百里香
rosemary /ˈrəʊzməri/ *n.* 迷迭香
mint /mɪnt/ *n.* 薄荷
nutmeg /ˈnʌtmeg/ *n.* 肉豆蔻
etc. abbr. /ɪtˈsetərə/ 等等
frequently used /ˈfriːkwəntli juːzd/ 常用的
cinnamon /ˈsɪnəmən/ *n.* 肉桂皮
bay leaf /beɪ liːf/ 月桂叶;香叶
clove /kləʊv/ *n.* 丁香
fennel seed /ˈfenl siːd/ 茴香籽
cumin /ˈkjuːmɪn/ *n.* 孜然
chili /ˈtʃɪli/ *n.* 辣椒
powder /ˈpaʊdə(r)/ *n.* 粉末
star anise /ˌstɑːr ˈænɪs/ 八角
5-spice /faɪv spaɪs/ 五香粉
Sichuan pepper /ˈsiːtʃwɑːn ˈpepə(r)/ 花椒

中文

Neil: 烹饪中使用的香草和香料有很多种，最常用的香料有……史蒂夫，请给我举几个例子。

Steve: 有罗勒、黑胡椒、百里香、迷迭香、薄荷、肉豆蔻等。我说的对吗？

Neil: 是的，你说对了。其中一些也经常用于中式烹饪，比如肉桂皮、月桂叶、丁香、茴香籽、孜然、辣椒粉这些。

Steve: 还有些是传统的中国香料，比如八角、五香粉和花椒。

Neil: 说得对。

注释

在厨房英语里，香料 (spice) 和香草 (herb) 这两个词通常结对出现。比如：What herbs and spices to add to the sauce？（酱汁里要加什么香草和香料？）它们都来自植物的某一部位，香料来自花瓣、种子、树皮或根茎等，香草来自植物的叶片。由于加工方式的不同，在烹饪领域，我们习惯将干燥过或者研磨过的调料称为香料，将叶片状（包括叶片被磨碎以后）的调料称为香草。

举例说明：

Herbs and Spices 香草及香料	
herbs 香草	**spices** 香料
basil 罗勒	clove 丁香
thyme 百里香	nutmeg 肉豆蔻
rosemary 迷迭香	cinnamon 肉桂皮
coriander 香菜；芫荽	star anise 八角
mint 薄荷	garlic 大蒜
bay leaf 香叶	black pepper 黑胡椒
parsley 欧芹	chili 辣椒
oregano 牛至叶	paprika 红辣椒粉

3. Adding Condiments 添加调味品

(初级厨师：Steve，西餐厨师长：Neil)

Neil: For the topping: chop onion, coriander, and garlic. Then mix together.

Steve: What condiments to put, chef?

Neil: For spice, add cinnamon, nutmeg, and paprika.

Steve: What's next?

Neil: Season with sugar, salt and pepper, and combine with olive oil.

词汇

topping /ˈtɒpɪŋ/ n. （菜肴、蛋糕等上的）浇汁，浇料，配料，佐料

onion /ˈʌnjən/ n. 洋葱

coriander /ˌkɒriˈændə(r)/ n. 香菜，芫荽

garlic /ˈɡɑːlɪk/ n. 大蒜，蒜头

cinnamon /ˈsɪnəmən/ n. 肉桂皮

nutmeg /ˈnʌtmeɡ/ n. 肉豆蔻

paprika /ˈpæprɪkə/ n. 红辣椒粉

season /ˈsiːzn/ v. 给……调味，加作料

sugar /ˈʃʊɡə(r)/ n. 糖

Steve: Yes, chef.
Neil: Onto the plate. Start placing it around the meat.
Steve: Shall I sprinkle with some fresh mint or rosemary?
Neil: With fresh mint. It's done.

---

salt /sɔːlt/ **n.** 盐
pepper /ˈpepə(r)/ **n.** 胡椒；辣椒
combine with /kəmˈbaɪn wɪð/ 与……结合
olive oil /ˌɒlɪvˈɔɪl/ 橄榄油
onto /ˈɒntu/ **prep.** 向……之上
placing /ˈpleɪsɪŋ/ **v.** 放置（place 的现在分词）
sprinkle /ˈsprɪŋk(ə)l/ **v.** 洒，撒
fresh mint /freʃ mɪnt/ 新鲜薄荷叶
rosemary /ˈrəʊzməri/ **n.** 迷迭香

中文

Neil： 做配料：将洋葱、香菜和大蒜切碎。然后混合在一起。
Steve： 厨师长，要放什么调味品？
Neil： 香料的话，加肉桂、肉豆蔻和辣椒粉。
Steve： 接下来呢？
Neil： 加糖、盐和胡椒调味，再加入橄榄油。
Steve： 好的，厨师长。
Neil： 上碟。开始把配料放在肉的周围。
Steve： 我要不要撒些新鲜薄荷叶或迷迭香？
Neil： 放新鲜薄荷叶。搞定。

注释

表达在食材上添加调味品 (condiments) 以进行调味 (season) 的句型
（1）加入：Add/put in _____.
　　Add/put in peanut oil. 加入花生油。
　　Add/put in chili powder. 加入辣椒粉。
（2）用_____调味：Season with _____.
　　Season with salt and pepper. 用盐和胡椒调味。
　　Season with spice. 用香料调味。
　　Season beef with black pepper and rosemary.
　　用黑胡椒和迷迭香给牛肉调味。
（3）一边加_____（调料）一边调和：Stir in _____.
　　Stir in the soy sauce. 加入酱油并搅拌。
　　Stir in cornstarch. 加入玉米淀粉搅拌。
（4）将_____与_____混合：Mix _____ with _____.
　　Mix salt with water. 把盐和水混合。
　　Mix with some sesame, ginger powder, and MSG. 混合一点芝麻、姜粉和味精。
（5）撒上：Sprinkle(with) _____. 或 Sprinkle _____ (over, on) _____.
　　Sprinkle with cheese powder./Sprinkle cheese powder on top. 撒上奶酪粉。
　　Sprinkle with basil./Sprinkle basil over the meat. 把罗勒撒在肉上。

Condiments and Seasonings			
oil /ɔɪl/	油	peanut oil /ˈpiːnʌt ɔɪl/	花生油
olive oil /ˈɒlɪv ɔɪl/	橄榄油	sesame oil /ˈsesəmi ɔɪl/	芝麻油
sunflower oil /ˈsʌnflaʊə(r) ɔɪl/	葵花籽油	lard /lɑːd/	猪油
cream /kriːm/	奶油	oyster sauce /ˈɔɪstə ˈsɔːs/	蚝油
soy sauce /ˌsɔɪ ˈsɔːs/	酱油	light soy sauce /laɪt sɔɪ sɔːs/	生抽
dark soy sauce /dɑːk sɔɪ sɔːs/	老抽	liquor /ˈlɪkə(r)/	酒，白酒
rice wine /raɪs waɪn/	米酒	cooking wine /ˈkʊkɪŋ waɪn/	料酒
vinegar /ˈvɪnɪɡə(r)/	醋	vinaigrette /ˌvɪnɪˈɡret/	油醋汁
mature vinegar /məˈtʃuə ˈvɪnɪɡə(r)/	陈醋	balsamic vinegar /bɔːlˌsæmɪk ˈvɪnɪɡə(r)/	香醋
salt /sɔːlt/	盐	sea salt /siː sɔːlt/	海盐
kosher salt /ˈkəʊʃə(r) sɔːlt/	粗盐；犹太盐	Monosodium Glutamate (MSG) /ˌmɒnəʊˈsəʊdɪəm ˈɡluːtəmeɪt/	味精
chicken essence /ˈtʃɪkɪn ˈesns/	鸡精	pepper /ˈpepə(r)/	胡椒
Sichuan pepper /ˈsiːtʃwɑːn ˈpepə(r)/	花椒	black pepper /ˌblæk ˈpepə(r)/	黑胡椒
sugar /ˈʃʊɡə(r)/	糖	brown sugar /ˌbraʊn ˈʃʊɡə(r)/	红糖
caramel /ˈkærəmel/	焦糖	yeast /jiːst/	酵母
curry /ˈkʌri/	咖喱	honey /ˈhʌni/	蜂蜜
sesame paste (sauce) /ˈsesəmi peɪst/	芝麻酱	chili sauce /ˈtʃɪli sɔːs/	辣椒酱
Tabasco /təˈbæskəʊ/	辣椒酱 塔巴斯科辣椒	starch /stɑːtʃ/	淀粉
flour /ˈflaʊə(r)/	面粉	wheat flour /wiːt ˈflaʊə(r)/	麦粉

（续）

Condiments and Seasonings

butter /ˈbʌtə(r)/	黄油	mustard /ˈmʌstəd/	芥末
ketchup /ˈketʃəp/	番茄酱	dressing /ˈdresɪŋ/	沙拉酱；淋汁
mayonnaise /ˌmeɪəˈneɪz/	蛋黄酱	herb /hɜːb/	草本植物，药草
spice /spaɪs/	香料	bay leaf /ˈbeɪ liːf/	香叶
vanilla /vəˈnɪlə/	香草	basil /ˈbæzl/	罗勒
Chinese basil /ˌtʃaɪˈniːz ˈbæzl/	紫苏	nutmeg /ˈnʌtmeg/	肉豆蔻
cumin /ˈkjuːmɪn/	孜然	thyme /taɪm/	百里香
cinnamon /ˈsɪnəmən/	肉桂皮	rosemary /ˈrəʊzməri/	迷迭香
saffron /ˈsæfrən/	藏红花	anise /ˈænɪs/	八角，茴香
dill /dɪl/	小茴香	mint /mɪnt/	薄荷
sage /seɪdʒ/	鼠尾草	wine /waɪn/	葡萄酒
syrup /ˈsɪrəp/	糖浆	lemon grass /ˈlemən grɑːs/	香茅草
turmeric /ˈtɜːmərɪk/	姜黄	clove /kləʊv/	丁香
vanilla extract /vəˈnɪlə ˈekstrækt/	香草精	chicken broth /ˈtʃɪkɪn brɒθ/	浓缩鸡汤（鸡汤块）
smoked paprika /sməʊkt ˈpæprɪkə/	红辣椒粉（匈牙利）	cayenne pepper /keɪˈen pepə(r)/	卡宴辣椒粉
Mozzarella cheese /ˌmɒtsəˈrelə tʃiːz/	马苏里拉奶酪	Parmesan cheese powder /ˌpɑːməˈzæn tʃiːz ˈpaʊdə(r)/	帕尔马干酪粉

练习：掌握常见的调味品词汇

完成以下食材翻译练习，并背诵词汇。

序号	中文	英文	序号	中文	英文	序号	中文	英文
1	酱油		13	盐		25	糖	
2	味精		14	料酒		26	醋	
3	焦糖		15	粗盐		27	咖喱	
4	辣椒酱		16	芥末		28	黄油	
5	鸡精		17	百里香		29	红辣椒粉	
6	罗勒		18	花生油		30	红酒	
7	蜂蜜		19	迷迭香		31	蚝油	
8	芝麻酱		20	橄榄油		32	胡椒	
9	薄荷		21	孜然		33	香叶	
10	香醋		22	花椒		34	肉桂皮	
11	八角		23	米酒		35	番茄酱	
12	糖浆		24	奶油		36	淀粉	

职业提示

人生百味，五味调和

酸甜苦辣咸是烹饪中的"五味"。厨师将酸甜苦辣咸这五种味道互相配伍、调和，把握适当的分寸，构成平衡、协调的味觉世界。主料、辅料和调料相辅相成、合而为一，调和之美是中国厨师长期以来遵循的最佳原则。五味调和与中国传统文化的"中庸之道"不谋而合。

Unit 16 Common Measure Words in the Recipe
食谱中常见的量词

在食谱里，表达调味品的数量时会涉及量词 (measure words)。中文与英文的量词用法有很大的不同，中文里表达数量的结构是"数词＋量词＋名词"，比如"一杯水""三个面包""五块黄油"；在英文里，很多情况下表达数量时可在可数名词前加上数词，比如"2 eggs"和"3 burgers"。不可数名词则使用量词词组表达，比如"a jar of milk"（一罐牛奶），"2 pounds of beef"（两磅牛肉），还常常用英文缩写表达——比如"8oz steak"（8盎司的牛肉）和"2c butter"（2杯黄油）等，外行人看了简直摸不着头脑。因此本单元特地讲解英文食谱里所涉及的一些常用量词词汇。

Learning Objectives 学习目标

❊ Learn the basic terms of cooking measurements and understand their abbreviations.
学习烹饪量词基础词汇并理解这些词汇的缩写。

❊ Learn to read a cooking recipe.
学会阅读烹饪食谱。

扫码看视频

Basic Terms 基础词汇

recipe 食谱 /ˈresəpi/	directions 做法 /dəˈrekʃns/	ingredients 材料 /ɪnˈɡriːdiənts/
gram /ɡræm/ 克 (g)	kilogram /ˈkɪləɡræm/ 千克 (kg)	ounce /aʊns/ 盎司 (oz)
litre /ˈliːtə(r)/ 升 (l)	millilitre /ˈmɪliliːtə(r)/ 毫升 (ml)	pound /paʊnd/ 磅 (lb)
teaspoon /ˈtiːspuːn/ 茶匙 (tsp)	tablespoon /ˈteɪblspuːn/ 汤匙 (tbsp)	inch /ɪntʃ/ 英寸 (in.)
cup /kʌp/ 杯	a tablespoon of /əˈteɪblspuːn əv/ 一汤匙	a teaspoon of /əˈtiːspuːn əv/ 一茶匙
a pinch of /ə pɪntʃ əv/ 少许	a touch of /ə tʌtʃ əv/ 少许	a drop of /ə drɒp əv/ 一滴

配料
/ɪnˈɡriːdɪənts/
ingredients

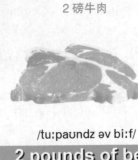
2磅牛肉
/tuː paʊndz əv biːf/
2 pounds of beef
2lb beef

2克盐
/tuː ɡræmz əv sɔːlt/
2 grams of salt
2g salt

茶匙
/ˈtiːspuːn/
teaspoon
tsp

一勺盐
/ə tiːspuːn əv sɔːlt/
a teaspoon of salt

杯
/kʌp/
cup

一杯米
/ə kʌp əv raɪs/
a cup of rice

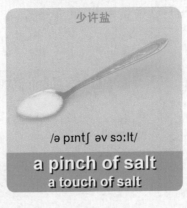
少许盐
/ə pɪntʃ əv sɔːlt/
a pinch of salt
a touch of salt

Functional Expressions 实用表达

请流利地朗读以下词组，并做到中英互译。

1. 30 克芝麻油　　　　30 grams of sesame oil
2. 30 克花生油　　　　30 g peanut oil
3. 10 克味精　　　　　10 g MSG
4. 200 克辣椒粉　　　 200 g chili powder
5. 1 公斤面粉　　　　 1 kg flour

gram /græm/ *n.* 克
g:gram 的缩写
kg:kilogram 的缩写

"10 g MSG" 读作：ten grams of MSG
"1kg flour" 读作：one kilogram of flour

6. 1 斤面粉   half a kilogram of flour
7. 1 勺盐   1 tsp salt
8. 2 勺糖   2 tsp sugar
9. 2 大勺橄榄油   2 tbsp olive oil
10. 6 盎司的比目鱼片   6 oz halibut fillet
11. 8 液盎司的原汤   8 fl oz stock

kilogram /ˈkɪləgræm/ *n.* 千克，公斤
teaspoon /ˈtiːspuːn/ *n.* 茶匙，一茶匙的量
"2 tbsp olive oil" 读作：two tablespoons of olive oil
"6 oz halibut fillet" 读作：six ounces of halibut fillet
ounces /ˈaʊnsɪz/ *n.* 盎司（复数）

"1 tsp salt" 读作：one teaspoon of salt
"2 tsp sugar" 读作：two teaspoons of sugar
tablespoon /ˈteɪblspuːn/ *n.* 大汤匙；一餐匙的量
ounce /aʊns/ *n.* 盎司
fl:fluid 的缩写

12. 0.5 磅的黄油   1/2 lb butter
13. 2 杯面粉   2 cup flour
14. 1/3 杯白葡萄酒   1/3 cup white wine
15. 1 升牛奶   1L milk
16. 10 毫升柠檬汁   10 ml lemon juice
17. 6 寸的奶酪蛋糕   6-inch cheesecake

"lb" 读作：pound of
pound /paʊnd/ *n.* 磅（重量单位，1 磅约合 0.454 千克）
"2 cup flour" 读作：two cups of flour
cup /kʌp/ *n.* 杯
litre /ˈliːtə(r)/ *n.* 升
millilitre /ˈmɪliliːtə(r)/ *n.* 毫升

"1/2 lb butter" 读作：half a pound of butter
"1/3 cup white wine" 读作：one third cup of white wine
"1 L milk" 读作：one litre of milk
"10 ml" 读作：10 millilitres of 或者 10 ml（/mɪlz/）of
inch /ɪntʃ/ *n.* 英寸

18. 2 小支新鲜的欧芹   2 sprigs of fresh parsley
19. 2 瓣蒜   2 cloves of garlic
20. 少许盐   A touch of salt
21. 一撮胡椒   A pinch of pepper
22. 一把香菜   A handful of coriander

sprig /sprɪɡ/ *n.* 带叶小枝
clove /kləʊv/ *n.* 蒜瓣；葱瓣
a touch of /ə tʌtʃ əv/ 少许，少量，有一点
pinch /pɪntʃ/ *n.* 捏，少量
handful /ˈhændfʊl/ *n.* 一把（的量）；用手抓起的数量

parsley /ˈpɑːsli/ *n.* 欧芹
garlic /ˈɡɑːlɪk/ *n.* 大蒜
a pinch of /ə pɪntʃ əv/ 一撮，少许
a handful of /ə ˈhændfʊl əv/ 一把

Task 任务   翻译练习

1. 中译英

腌料：生抽 2 茶匙、料酒 1 茶匙、糖 1 茶匙、玉米淀粉 1 茶匙、胡椒粉少许、麻油少许。（腌料 marinade，生抽 light soy sauce，料酒 cooking wine，糖 sugar，玉米淀粉 cornstarch，胡椒粉 pepper，麻油 sesame oil）

_____

_____

2. 英译中

Ingredients:
    5 lb lamb shanks
    4 finely chopped sprigs of parsley
    6 large potatoes (peeled and sliced)
    2 tbsp olive oil
    1 tsp salt
    1 pinch freshly ground black pepper
    3 cloves of garlic
    1 ml vanilla extract

💬 Dialogues 对话

1. To Make Chicken Stock 炖鸡汤

（初级厨师：Steve，中餐厨师长：David）

David: Go get a big stockpot, I'm going to make chicken stock with a whole chicken.

Steve: Here is a 5-litre stockpot.

David: Big enough. Then prepare 300g old ginger, 20ml cooking wine, and 2.5l mineral water.

Steve: Yes, chef.

David: When the chicken stock is ready to use, it is a great ingredient for soups, noodles, wonton, and other dishes.

词汇

chicken stock /ˈtʃɪkɪn stɒk/ 鸡汤
stock /stɒk/ n. 高汤；原汤
stockpot /ˈstɒkpɒt/ n. 汤锅
whole chicken /həʊl ˈtʃɪkɪn/ 整鸡，全鸡
5-litre /faɪv ˈliːtə(r)/ 五升的
300 g 读作：300 grams，意为 300 克
gram /græm/ n. 克
20ml 读作：20 millilitres，意为 20 毫升
millilitre /ˈmɪlɪliːtə(r)/ n. 毫升
2.5l 读作：2.5 litres，意为 2.5 升
litre /ˈliːtə(r)/ n. 升
mineral water /ˈmɪnərəl ˈwɔːtə(r)/ 矿泉水
wonton /ˌwɒnˈtɒn/ n. 馄饨

中文

David: 拿一个大汤锅，我要用一整只鸡做鸡汤。
Steve: 这是一个 5 升的汤锅。
David: 够大了。然后准备好老姜 300 克，料酒 20 毫升，矿泉水 2.5 升。
Steve: 好的，厨师长。
David: 当鸡汤煮好后，它是制作靓汤、面条、馄饨和其他菜肴的好原料。

注释

英文量词的书面语与口语在表达时需要注意：
300 g old ginger 读作：three hundred grams of old ginger
20 ml cooking wine 读作：twenty millilitres of cooking wine 或者 twenty ml(读作：/mɪlz/) of cooking wine
2.5 l mineral water 读作：two point five litres of mineral water
注：英文里量词词组是单数时：a+（量词的单数形式）of（名词）；是复数时：数词+（量词的复数形式）of（名词）。

2. Re-Seasoning 重新调味

（初级厨师：Steve，行政副厨师长：Jackson）

Jackson: Taste the juice, Steve. It's tasteless.
Steve: I have added 1 teaspoon of salt to it.
Jackson: Just correct the seasoning. Now add more salt and pepper. Add a pinch of sugar and a few drops of vinegar.
Steve: It's savoury now.
Jackson: Check seasoning at different stages of cooking.
Steve: I'm always afraid of oversalting the dish.
Jackson: Follow the recipe and use right amount of salt. If a recipe simply says "salt to taste", just keep in mind that use 2 teaspoons per pound for meat dish.
Steve: Thanks for your kind reminder, chef.

词汇

juice /dʒuːs/ **n.** 肉汁
tasteless /ˈteɪstləs/ **adj.** 无味的
teaspoon /ˈtiːspuːn/ **n.** 茶匙，一茶匙的量
correct /kəˈrekt/ **v.** 纠正
a pinch of /ə pɪntʃ ɒv/ 少许，一撮
pinch /pɪntʃ/ **n.** 少量；捏，掐
drop /drɒp/ **n.** 少量，滴；水珠
savoury /ˈseɪvəri/ 咸味的，咸口的；好吃的
oversalt /ˈəʊvə(r)sɔːlt/ **v.** 放过量的盐
oversalting /ˈəʊvə(r)sɔːltɪŋ/ **n.** 加盐过多
amount /əˈmaʊnt/ **n.** 数量
salt to taste /sɔːlt tu: teɪst/ 根据口味放盐
per pound /pə(r) paʊnd/ 每磅
per /pə(r)/ **prep.** 每；每一
pound /paʊnd/ **n.** 磅（重量单位，1 磅约合 0.454 千克）
reminder /rɪˈmaɪndə(r)/ **n.** 提醒

中文

Jackson： 尝一尝肉汁，史蒂夫。没有味道。
Steve： 我加了 1 茶匙盐了。
Jackson： 纠正调味就好。现在加多点盐和胡椒粉调味。加少许糖和几滴醋。
Steve： 现在味道好了。
Jackson： 在烹饪的不同阶段要检查调味。
Steve： 我老是担心菜咸了。
Jackson： 遵循食谱并使用适量的盐。如果食谱中只说"按照口味加盐"，则请记住，每磅肉菜使用 2 茶匙的盐。
Steve： 感谢您的提醒，厨师长。

注释

（1）在阅读英文食谱（recipe）时，我们需要了解汤匙（tablespoon）和茶匙（teaspoon）的英文缩写及标准容积。

tbsp/T = tablespoon   tsp/t = teaspoon

以一套六件的量勺套装（6-piece measuring spoon set）为例，它们的缩写及对应的容积分别是：

1/8 tsp = 0.63 ml   1/4 tsp = 1.25 ml
1/2 tsp = 2.5 ml   1 tsp = 5 ml
1/2 tbsp = 7.5 ml   1 tbsp = 15 ml

注：量勺是按体积而不是重量来使用，不同食材因密度不同，重量会有差别：1 汤勺的酱油重量约为 15 克，而 1 汤勺的花生油重量约为 8 克。

（2）在表达所需的数量时，注意分数的读法：分子是基础词，分母是序数词；分子是 1 以上的任何数时，作分母的序数词要用复数形式。

分子：基数词（one，two，three…）
分母：序数词（first，second，third…）

1）1/8 读作：a/one eighth
2）1/4 读作：a/one quarter 或者 a/one fourth
3）1/2 读作：a/one half
4）3/4 读作：three quarters 或者 three fourths

（3）翻译"少许"可用的量词词组有 a touch of，a pinch of 以及 a splash of。如：

a little pinch/ touch of sugar 少许糖   a pinch/touch of salt 少许盐
a splash of olive oil 少许橄榄油

3. Making the Dough 做面团

(初级厨师：Steve，饼房领班：Rosa)

Rosa: Prepare all ingredients: 500 grams of flour, 7 grams of yeast, 10 grams of salt, and 3 tablespoons of olive oil.
Steve: How much water to add?
Rosa: About 300 ml of water. Use a measuring cup. At this stage, keep the salt away from yeast.
Steve: OK. Shall I begin to mix these dry ingredients now?
Rosa: Yes. Then add water. It'll form into a paste.
Steve: My fingers are like a mixer. I need to start kneading.
Rosa: Good. Keep working on the dough with your hands.

词汇

dough /dəʊ/ **n.** 生面团
grams /græmz/ **n.** 克（gram 的复数形式）
flour /ˈflaʊə(r)/ **n.** 面粉
yeast /jiːst/ **n.** 酵母
tablespoons /ˈteɪblspuːnz/ **n.** 大汤匙，大调羹（tablespoon 的复数）
ml = millilitre /mɪl/
measuring cup /ˈmeʒərɪŋ kʌp/ 量杯
form /fɔːm/ **v.** 形成
paste /peɪst/ **n.** 面糊，面团
finger /ˈfɪŋɡə(r)/ **n.** 手指
mixer /ˈmɪksə(r)/ **n.** 搅拌器
kneading /ˈniːdɪŋ/ **n.** 揉（面团等）；捏合

中文

Rosa：准备所有原料：500 克面粉、7 克酵母、10 克盐和 3 汤匙橄榄油。
Steve：要加多少水？
Rosa：约 300 毫升水。要使用量杯。在这个阶段，盐不要接触到酵母。
Steve：好。我现在就开始混合这些干的原料吗？
Rosa：是的。然后加水。它会变成糊状。
Steve：我的手指就像一台搅拌机。我要开始揉面了。
Rosa：好。继续用手揉面团。

注释

（1）中文里的"一斤"有两种英文表达。

1）half a kilogram：500 克，即一公斤（kilogram）的一半，这是国际通用的表达。

　　half a kilogram of grapefruit 一斤葡萄柚 /500 克葡萄柚
　　a half kilogram of pork 一斤猪肉 /500 克猪肉
　　How much for a half kilogram? 一斤多少钱？

2）catty /ˈkæti/：这是马来西亚和新加坡等东南亚国家常用的重量单位"斤"所对应的英文，使用范围不广。

　　a catty fruit 一斤水果
　　one catty Chinese kale 一斤芥兰

（2）毫升"millilitre"的缩写是"ml"，可读为"/ˈmɪliliːtə(r)/"或者"/mɪl/"。复数时，300ml 则读作：three hundred millilitres 或者 three hundred ml（读作：/mɪlz/）。

300ml of water (300 毫升水) 读作：
1) three hundred millilitres of water
2) three hundred ml of water

（3）常见的烹饪量词。

Measure Words			
gram (g) /græm/	克	50g eggplant 50 克茄子	fifty grams of eggplant
kilogram (kg) /ˈkɪləgræm/	千克	1kg wheat flour 1 公斤小麦粉	one kilogram of wheat flour
ounce (oz) /aʊns/	盎司	6 oz fillet steak 6 盎司的菲力牛排	six ounces of fillet steak
fluid ounces (fl oz) /ˌfluːɪd ˈaʊns/	液盎司	8 fl oz stock 8 液盎司的原汤	eight fluid ounces of stock
inch (in.) /ɪntʃ/	英寸	6-inch cake 6 寸的蛋糕	six-inch cake
pound (lb) /paʊnd/	磅	2 lb beef 2 磅的牛肉	two pounds of beef
teaspoon (tsp) /ˈtiːspuːn/	茶匙	1/2 tsp salt 1/2 茶匙盐	a half teaspoon of salt
tablespoon (tbsp) /ˈteɪblspuːn/	汤匙	2 tbsp of olive oil 两汤匙橄榄油	two tablespoons of olive oil
litre (L) /ˈliːtə(r)/	升	2L milk 2 升牛奶	two litres of milk
millilitre (ml) /ˈmɪliliːtə(r)/	毫升	2ml lemon juice 2 毫升柠檬汁	two millilitres of lemon juice
cup (c) /kʌp/	杯	1 cup rice 一杯米	one cup of rice

练习：翻译食谱成分

根据前文所提供的常见量词，将下表翻译为英文。

1. Ingredients for Grilled Rib-eye Steak 烤肋眼牛排的材料	
2 tsp minced fresh garlic	
2 tsp coarsely ground black pepper	
1 tsp dry mustard	
2 tsp paprika	
2 tsp chili powder	

（续）

1. Ingredients for Grilled Rib-eye Steak 烤肋眼牛排的材料	
1 tsp dried thyme leaves, crumbled	
1/2 tsp salt	
1 tbsp olive oil	
8 oz rib-eye steak	
2. Ingredients of Steamed Fish 蒸鱼的材料	
1 whole fish, gutted and scaled	
2 cloves garlic, chopped	
1 scallion, julienned	
1 medium ginger, peeled and julienned	
1 tbsp canola oil	
1 tbsp rice wine	
1/4 cup soy sauce	
crushed red pepper to taste	
1 pinch salt, or more to taste	
fresh cilantro sprigs, for garnish	

职业提示

诚信经营，用心服务

"言不信者，行不果。"古往今来，诚信是华夏文明的传统美德，是中华文化的重要价值观，也是商业成功的基石。这与党的二十大提出的建设社会主义现代化强国的目标不谋而合。我们要树立文明经商、诚实守信的经营理念。只有以道德经营为中心，才能在市场竞争中立于不败之地，取得更大的成功和发展。而"诚实交易，童叟无欺"就是"老字号"餐饮企业的立身之本、价值观念。中国全聚德集团从开业时起，它的创始人就创立了"聚拢德行"的经营理念。全聚德展厅里的"德"字，让人为之动容，而老店精湛的技艺和以德立店的文化价值更受到世人的青睐。

Unit 17　Common Functional Expressions in the Kitchen
厨房常用的功能句

　　厨房的工作需要多动手、少动嘴，团队协作力求高效。因此在表达各类食材的处理及工作内容时，使用的句型较为简单，以祈使句居多。省略了主语"you"后，以动词原型做句首，比如"bone the chicken"（将鸡肉剔骨），"scale the fish"（将鱼去鳞）；还涉及较多的动词，比如"drain"（沥干），"freeze"（冰冻），"pour in"（倒入），"add"（加入），"turn on"（打开）等，多看多练才能活用。

Learning Objectives 学习目标

❋ Learn basic verbs, phrases, and sentences of food materials processing.
　学习处理食品的基本单词、短语和句子。
❋ Able to make short dialogues about food materials processing in the kitchen.
　能在厨房进行关于食品处理的简短对话。

扫码看视频

Basic Terms 基础词汇

rinse 清洗 /rɪns/	drain 沥干 /dreɪn/	cook 煮 /kʊk/
cut 切 /kʌt/	grind 磨碎 /graɪnd/	chop 剁碎 /trɒp/
stir 搅拌 /stɜː(r)/	gut 去内脏 /gʌt/	bone 去骨 /bəʊn/
seed 去籽 /siːd/	peel 去皮 /piːl/	mix 混合 /mɪks/
sprinkle 撒上 /ˈsprɪŋk(ə)l/	pour in 倒入 /pɔː(r) ɪn/	add 加入 /æd/
turn on 开火 /tɜːn ɒn/	heat 加热 /hiːt/	cover 盖上盖 /ˈkʌvə(r)/
turn down 调小 /tɜːn daʊn/	garnish 装饰 /ˈgɑːnɪʃ/	serve 上菜 /sɜːv/

Functional Expressions 实用表达

请流利地朗读以下句子,并做到中英互译。

1. 沥干水。 Drain it/them.
2. 将食物冷冻。 Freeze the food.
3. 冷藏两个小时。 Refrigerate for 2 hours.
4. 把它放进冰箱。 Put it in the fridge/refrigerator/freezer/ cooler.
5. 将鸡肉放在冰箱内解冻一晚。 Defrost chicken overnight in the fridge.
6. 切蘑菇 / 洋葱 / 西瓜。 Cut the mushroom/onion/watermelon.
7. 把肉切成小块 / 丁块 / 薄片 / 条。 Cut the meat into pieces/cubes/slices/strips.
8. 把蒜头剁碎。 Finely chop the garlic.
9. 磨黑胡椒。 Grind some black pepper.
10. 把鸡去骨。 Bone the chicken.

drain /dreɪn/ v. 沥干
refrigerate /rɪˈfrɪdʒəreɪt/ v. 使冷却
refrigerator /rɪˈfrɪdʒəreɪtə(r)/ n. 电冰箱(同时有制冰及冷藏功能)
freezer /ˈfriːzər/ n. 冰柜,冷冻库
注:"冰箱"的英文对应有"fridge/refrigerator/freezer/cooler",注意区分。
defrost /diːˈfrɒst/ v. 解冻
cut /kʌt/ v. 切
pieces /ˈpiːsɪz/ n. 块,片,段(piece 的复数)
slices /ˈslaɪsɪz/ n. 切片(slice 的复数)
finely chop /ˈfaɪnli tʃɒp/ 剁碎,细细地切
bone /bəʊn/ v. 剔去……的骨; n. 骨头

freeze /friːz/ v. 使冻结
fridge /frɪdʒ/ n. 电冰箱
cooler /ˈkuːlə(r)/ n. 冷饮机

overnight /ˌəʊvəˈnaɪt/ adv. 一夜之间;过一个晚上
cut...into... /kʌt ˈɪntə/ 切成……(形状)
cubes /kjuːbz/ n. 立方体(cube 的复数)
strips /strɪps/ n. 条
grind /graɪnd/ v. 磨碎

11. 把鱼去鳞。 Scale the fish.
12. 把鱼去内脏。 Gut the fish.
13. 将土豆去皮。 Peel potatoes.
14. 将南瓜去籽。 Remove pumpkin seeds.
15. 切成两半。 Cut in half.
16. 切成四份。 Cut into fourths.
17. 打两个鸡蛋,然后搅打。 Break 2 eggs and then whisk.
18. 把它们混合在一起。 Mix them together.
19. 和牛奶混合。 Blend it with milk.
20. 放在一边。 Set it aside.

scale /skeɪl/ v. 刮鳞; n. 鳞
remove /rɪˈmuːv/ v. 移走,拿开
break /breɪk/ v. 打破
blend with /blend wɪð/ 与……混合

gut /gʌt/ v. 取出内脏
seed /siːd/ n. 籽;种子
whisk /wɪsk/ v. 搅拌,搅动
aside /əˈsaɪd/ adv. 在旁边;撇开

peel /piːl/ v. 去皮; n. 皮
fourth /fɔːθ/ n. 第四;四分之一
mix together /mɪks təˈgeðə(r)/ 混合

21. 撒上盐。	Sprinkle some salt.
22. 涂上黄油。	Coat with butter.
23. 倒入酱油。	Pour in some soy sauce.
24. 加入葱。	Add in scallion.
25. 放入花生油。	Put in some peanut oil.
26. 点着炉灶。	Turn on the stove.
27. 关火。	Turn off the fire.
28. 关煤气。	Turn off the gas.
29. 热锅。	Heat the pan.
30. 调大火量。	Turn up the flame/raise the heat.

sprinkle /ˈsprɪŋk(ə)l/ v. 洒，撒　　coat /kəʊt/ v. 给……涂上一层　　turn on /tɜːn ɒn/ 打开；发动
stove /stəʊv/ n. 火炉　　turn off /tɜːn ɒf/ 关闭，关掉　　fire /ˈfaɪə(r)/ n. 火；炉火
gas /gæs/ n. 煤气；天然气；气体　　heat /hiːt/ v. 把……加热 n. 高温；温度
flame /fleɪm/ n. 火焰；火舌　　raise /reɪz/ v. 提高；升起

31. 调小火量。	Turn down the flame/ reduce the heat.
32. 用小火煮。	Over low heat.
33. 用中火煎。	Fry over medium heat.
34. 用大火翻炒。	Stir-fry over high heat.
35. 用文火煨。	Simmer it slowly.
36. 烧开。	Bring it to a boil.
37. 盖上锅盖。	Cover the pan.
38. 开盖。	Uncover the pan.
39. 根据口味加盐和胡椒粉。	Add salt and pepper to taste.
40. 用罗勒叶装饰。	Garnish with basil leaves.

reduce /rɪˈdjuːs/ v. 减少；降低　　low heat /ləʊ hiːt/ 小火　　medium heat /ˈmiːdiəm hiːt/ 中火
high heat /haɪ hiːt/ 大火　　boil /bɔɪl/ n. 沸腾，沸点　　cover /ˈkʌvə(r)/ v. 覆盖
uncover /ʌnˈkʌvə(r)/ v. 掀开盖子　　taste /teɪst/ n. 味道　　garnish /ˈgɑːnɪʃ/ v. 加装饰；加饰菜于

Task 任务　　翻译练习

1. 试着用简单的英语告诉同桌如何做一份美味的单面煎蛋（sunny side up fried egg）。
   比如：热锅、用大火煎、打鸡蛋、调小火、关火、放盐等。

2. 试着用简单的英语告诉同桌如何做一份辣椒炒肉片（fried sliced pork with chilli pepper）。
   比如：切肉片与切辣椒丝、热锅、倒油、先大火炒肉片、放调料、加入辣椒、调小火、关火等。

参考表达：

Common kitchen function sentences
常见厨房功能句

rinse and drain 清洗和沥干 rinse the lettuce and drain it 把生菜洗净沥干	peel 去皮 peel the garlic 大蒜去皮 bone 去骨 bone the chicken 鸡肉剔骨 scale 去鳞 gut 去内脏	mix together 混合 blend 搅拌 blend it with milk 和牛奶混合 *blend A with B whisk 搅拌（蛋） break 破开	cut into... 切成...... cut in half 切成两半 cut into fourths 切成四份 cut into pieces/cubes/slices/strips 切成小块 / 丁块 / 薄片 / 条	tear 撕开 cut 切 slice 切片 chop up 垛；切细 tear/cut/slice/chop up the lettuce 将生菜撕开、切开、切片、切成小块
finely chop 剁碎 finely chop the garlic 把蒜头剁碎 grind 磨碎 grind some black pepper 磨碎黑胡椒	add 加上 add in 加入 put in 放入 stir in 搅拌而加入 pour in 倒入	sprinkle 撒 sprinkle some salt 撒盐 to taste 根据口味 add salt and pepper to taste 根据口味加盐和胡椒粉	coat 涂上 coat the turkey with butter 将火鸡抹上黄油 set it aside 静置 set it aside for a few minutes 把它放在一边静置几分钟	cover the pan 合上锅盖 uncover the pan 掀开锅盖 *pan 平底锅；pot 锅 remove 移开、移走 remove the seed 去籽 remove the pan 起锅
turn on 打开 turn off 关闭 turn on the stove 点着炉灶 turn off the fire 关火 turn off the flame 关火 turn off the gas 关煤气	heat 热量；温度 flame 火焰 fire 火；炉火	low heat 小火 stir-fry over low heat 小火翻炒 medium heat 中火 over medium heat 用中火煮 high heat 大火 over high heat 用大火煮	turn down the flame 调小火量 reduce the heat 调小火量 turn up the flame 调高火量 raise the heat 调高火量	bring it to a boil 烧开 bring the soup to a boil 把汤加热至沸腾 simmer it slowly 用文火煨
heat 加热 heat the oil 热油 heat the butter 热黄油 warm up 加热 warm up the coffee 把咖啡加热	let it cool 冷却 let the soup cool 让汤冷却 freeze 冷冻 freeze the meat 把肉冷冻 refrigerate 冷藏 refrigerate for 2 hours 冷藏两个小时	turn brown 变棕色 until it turns brown 直到变成棕色 become soft 变软 flavor comes out 散发香味 until the flavor comes out 直到散发味道出来	onto plate 装碟 serve carrots onto plates 把胡萝卜装到这些盘子里 serve 上菜 ready to serve 可以上菜了 presentation 摆盘 a great presentation 很棒的摆盘	garnish 装饰 garnish with basil leaves 用罗勒叶装饰 decorate 装饰（通常不可吃） decorate the dish with dill and thyme 用莳萝和百里香装饰这道菜

 Dialogues 对话

1. To Stir-Fry Chinese Chives with Eggs 做韭菜煎鸡蛋

（初级厨师：Steve，中餐厨师长：David)

David: Rince the Chinese Chive and drain it with a colander. Then chop up.
Steve: Then fry the eggs?
David: No. Now crack 3 eggs, then blend with the finely chopped Chinese Chive. Whisk with chopsticks.
Steve: OK. No seasoning?
David: Add salt, Shaoxing wine and water. Let's start. Heat the pan. Pour in some peanut oil.
Steve: Over medium heat or high heat?
David: Over high heat. Stir in the egg mixture slowly. When eggs turn brown, it's cooked.
Steve: Smells good. Simple but very tasty.

词汇
rinse /rɪns/ v. 冲洗
drain /dreɪn/ v. 沥干
colander /ˈkʌləndə(r)/ n. 滤器
chop up /tʃɒp ʌp/ 切碎；切细；剁；切开
blend /blend/ v. 混合
finely chopped /ˈfaɪnli tʃɒpt/ 切得细碎的
whisk /wɪsk/ v. 搅动；n. 打蛋器
chopsticks /ˈtʃɒpstɪks/ n. 筷子
mixture /ˈmɪkstʃə(r)/ n. 混合物
egg mixture /eg ˈmɪkstʃə(r)/ 鸡蛋混合物；蛋液
cooked /kukt/ adj. 煮熟的

中文

David：冲洗韭菜，用沥水篮沥干水。然后切碎。

Steve：接着煎鸡蛋吗？

David：不。现在打3个鸡蛋，然后加入切碎的韭菜。用筷子搅拌。

Steve：好的。不用调料吗？

David：加盐、绍兴酒和水。我们开始吧。热锅。倒一些花生油。

Steve：用中火还是大火？

David：大火。倒入蛋液搅拌。当鸡蛋变成褐色时就熟了。

Steve：好香。简单但是很美味。

注释

（1）"rinse"和"wash"都指用水洗，区别是rinse强调快速冲洗且不加任何洗涤产品，在英语里对食材的清洗更多使用"rinse"而不是"wash"。

比如：rinse with warm water

用温水冲洗

rinse the leaves thoroughly under running water

用自来水把叶子彻底冲洗干净

（2）和处理鸡蛋有关的常用表达。

打鸡蛋（复数）：break the eggs/crack the eggs

打蛋器：whisk

搅拌鸡蛋（复数）：whisk the eggs

蛋黄：egg yolk　　蛋白：egg white　　蛋壳：eggshell

蛋液：egg mixture

蛋花：egg drop

用打蛋器打蛋：beat the eggs with a whisk

将蛋黄与蛋白分离：separate the egg yolk from the egg white

将鸡蛋与面粉混合：mix eggs with flour

蓬松的鸡蛋：fluffy eggs

蛋黄可流动的溏心蛋：runny eggs

* 复习Unit 5关于蛋的熟度的表达：easy用来形容煮得嫩的鸡蛋，包括煎蛋、荷包蛋和煎蛋卷；medium用来形容煮得半熟的鸡蛋；hard用来形容全熟的鸡蛋；soft形容煮得嫩的水煮蛋。

2. Preparing the Ingredients 准备食材

（初级厨师：Steve，西餐厨师长：Neil)

Neil: Please cut the mushrooms and tomatoes in half, and prepare a few sprigs of thyme for garnish.
Steve: Yes, chef.
Neil: And grind some dried red peppers into flakes with a mortar.
Steve: Shall I grind a small amount of Sichuan pepper into powder as well?
Neil: Yes, go ahead.

词汇

cut...in half /kʌt ɪn hɑːf/ 对半切，切成两半
sprig /sprɪɡ/ n. 小枝；带叶小枝
sprigs of... /sprɪɡz əv/ 几支，几条……
garnish /ˈɡɑːnɪʃ/ n.（食物）的装饰；配菜
grind /ɡraɪnd/ v. 磨碎
dried /draɪd/ adj. 干的
flake /fleɪk/ n. 小薄片，碎片
mortar /ˈmɔːtə(r)/ n. 钵，研钵，臼
a small amount of /ə smɔːl əˈmaʊnt əv/ 少量的
powder /ˈpaʊdə(r)/ n. 粉；粉末

中文

Neil: 请把蘑菇和番茄切成两半，再准备几枝新鲜的百里香做装饰。
Steve: 好的，厨师长。
Neil: 再用石钵把干辣椒磨成碎片。
Steve: 我要顺便把少量花椒磨成粉末吗？
Neil: 好，去做吧。

注释

（1）使用"cut...into"来表达"切成……的形状"时。

切成丁	cut into cubes/dices
切成小块	cut into pieces
切成大块	cut into chunks
切成片	cut into slices
切成条（丝）	cut into julienne
切两半	cut in half
切成四份	cut into fourths/quarters

* 复习 Unit 11 第 2 个对话 "The Knife Skills 刀工" 里关于食物不同形状相关的表达。

（2）和 "grind" 有关的词汇。

grinder	研磨机
meat grinder	绞肉机
coffee grinder	咖啡研磨机

*grind 的过去分词为 ground，指 "磨细的；磨碎的"，比如：

| ground beef | 绞碎的牛肉 |
| ground ginger | 姜粉 |

（3）"flake"一词在食物英语里有"把（鱼、食物等）切成薄片"以及"薄片；脆片饼干"之意。比如：

corn flakes	玉米片
red pepper flakes	干红辣椒碎
flaked salmon	干三文鱼肉片
flake the cod	把鳕鱼切成薄片

（4）"a small amount of..."翻译为"少量的……"，一般接不可数名词；"大量的……"则是"a large amount of..."。

（5）"mortar"指钵，配套的杵（槌）则是"pestle"（/ˈpesl/），"mortar and pestle"通常在句子里一起出现。如：

Use a mortar and pestle to crush the seed. 用研钵和杵把种子压碎。

3. Food Plating 摆盘

(初级厨师：Steve，西餐厨师长：Neil)

Neil: The noodles are just plated. Now use some edible materials to decorate the dish. Get creative!

Steve: Emm, herbs are common, flowers don't seem to fit it.

Neil: Choose whatever we have on hand. Art comes from life.

Steve: How about plating with sauces? Draw with chili or pesto sauces on the plate.

Neil: As far as it's well-plated. Start developing your food presentation skills.

Steve: I'd also like to use a few seared shrimps and a touch of pepper salt for garnish.

Neil: Just do it. It should be a beautifully plated dish.

词汇

plate /pleɪt/ v. 装碟，摆盘（过去分词 plated /ˈpleɪtɪd/）
edible /ˈedəb(ə)l/ adj. 可食用的
materials /məˈtɪəriəlz/ n. 材料
decorate /ˈdekəreɪt/ v. 装饰，装点
creative /kriˈeɪtɪv/ adj. 有创造力的，有想象力的
herbs /hɜːbs/ n. 草药，香草（herb 的复数）
common /ˈkɒmən/ adj. 普遍的，一般的
fit /fɪt/ v. 适合；和……相称
pesto /ˈpestəʊ/ n. 意大利青酱，香蒜酱
well-plated /wel ˈpleɪtɪd/ 摆得好的
developing /dɪˈveləpɪŋ/ v. 发展；研发
presentation /ˌpreznˈteɪʃn/ n.（这里指）菜肴的呈现方式
skills /ˈskɪlz/ n. 技能，技巧
seared /sɪrd/ adj. 炙烤的，炙烧的
beautifully /ˈbjuːtɪfli/ adv. 漂亮地；美好地

中文

Neil: 面条装好碟了。用一些可食用的材料来装饰。发挥创意！
Steve: 嗯，草叶类很常见，花似乎不适合。
Neil: 我们手头有什么就选什么。艺术源于生活。
Steve: 用调味汁来装饰怎么样？在盘子上涂辣椒酱或香蒜酱。

Neil: 只要摆得好就行。开始培养你的摆盘技巧。
Steve: 我还想用几只烤虾和一点椒盐来摆盘。
Neil: 用吧。它应该是一道漂亮的菜。

注释

（1）厨房英语里，"plate"做动词时可翻译为"装碟"，被动形式"plated"理解为"摆好盘的"。比如：

plate a fish dish　给一道鱼菜装碟
the fish is plated　鱼肉已经装好碟了

＊"摆盘"一词可对应"plating"（名词）以及"presentation"（名词）：

food plating/presentation　　　　食物摆盘（"plating"一词更倾向于表达"把食物在碟子上摆好"之意）

plating/presentation techniques　摆盘技巧
exquisite presentation　　　　　　精致的摆盘
beautiful plating　　　　　　　　　漂亮的装碟

（2）摆盘时所使用的材料在英语里为"garnish"，译为"配菜，装饰菜"，使用的材料绝大多数是可食用的。举例说明如下：

use parsley as garnish　　　　　　用欧芹做配菜
use a sprig of thyme for garnish　用一小枝百里香做装饰
add a garnish to this dish　　　　给这道菜加点装饰

"garnish"也做动词用：

garnish with basil leaves　　　　 用罗勒叶装饰
garnish the dish with dill　　　　用莳萝装饰这道菜

＊"decoration"所对应的摆盘装饰材料不一定是可食用的，比如生食材、花瓣、蛋糕上的卡片等。

若同学们对于单词分别作为名词和动词时所在句型位置，以及究竟要搭配什么介词（as, with, for 等）产生疑惑，可从英语里简单句的五种基本句型入手，理清句子成分及词性。将本文的实用英语知识作为输入，积累了足够的输入后才能渐进渐悟。

练习：翻译处理及制作食物时常见的表达

试着根据提示的词汇将句子翻译为英文。

1. 在碗里倒入鸡蛋。 倒入 pour in

2. 把生菜洗干净。 清洗 rinse；生菜 lettuce

3. 将整只鸭涂上腌料。 腌料 marinade；涂上 coat

4. 撒上盐和胡椒粉。 撒 sprinkle

5. 开火。 打开 turn on

6. 热花生油。 加热 heat

7. 大火翻炒。 大火 high heat；翻炒 stir-fry

8. 盖上锅盖，调小火。 盖上 cover；调小 reduce

9. 掀开锅盖，根据口味加盐。 开盖 uncover；根据口味 to taste

10. 用香菜做装饰。 装饰 garnish

11. 用削皮刀削土豆。 削皮 peel；削皮刀 peeler

12. 把土豆切成小块。 切成小块 cut into pieces

13. 把姜切成细丝。 切成丝 cut...into fine jullienne

14. 南瓜去籽。 去籽 remove the seeds

15. 把干辣椒磨碎。 磨碎 grind；干辣椒 dried red peppers

16. 用钵和杵捣碎大蒜。 捣碎 crush；钵和杵 mortar and pestle

17. 把汤烧开。 把……烧开 bring...to a boil

18. 把食材放进烤箱。 食材 ingredients；烤箱 oven

19. 倒入两勺酱油。 两勺 2 teaspoons of；酱油 soy sauce

20. 用淀粉勾芡。 淀粉 starch；勾芡 thicken

职业提示

工匠精神，个人意志

"复杂的事情简单做，简单的事情重复做。"要想成为一名受人称道的大厨，获得客人的认可，唯有一丝不苟、精心打磨自己的出品。厨师们做菜时禀怀着强烈的责任感，将自己的情感赋予手中的食材、菜肴，吸收先进的技艺，反复实验、创新，专注于提高自己的厨艺，用真情传递生命的温度，才能引发客人的内心共鸣。习近平总书记曾在一次座谈会上说，"一个国家文化的魅力、一个民族的凝聚力，主要通过语言表达和传递"。语言是传递饮食文化的重要途径，要不断学习和探索，通过你我的语言表达和传递中华饮食文化的魅力。

Pronunciation and Phonetic Symbols
读音和音标

元音和双元音 Vowels and diphthongs（23 个）					
iː	i	ɪ	e	æ	ɑː
ɒ	ɔː	ʊ	u	uː	ʌ
ɜː	ə	ɪə	əʊ	oʊ	aɪ
aʊ	ɔɪ	eɪ	eə	ʊə	
辅音 Consonants（24 个）					
p	b	t	d	k	g
tʃ	dʒ	f	v	θ	ð
s	z	ʃ	ʒ	h	m
n	ŋ	l	r	j	w

注：1. 表中的注音字符被称为 DJ 音标、Jones 音标或 EPD 音标。EPD 指由丹尼尔·琼斯（Daniel Jones）编写的《英语发音词典》（English Pronouncing Dictionary）一书，该书于 2011 年已修订至第 18 版。
2. 此表参考自《牛津高阶英汉双解词典（第 9 版）》（商务印书馆，2018 年）的"附录 9　读音和音标"（Appendix 9 Pronunciation and Phonetic Symbols）。
3. 可搜索微信小程序"BBC 国际音标"及"英语国际音标"等在线点读工具以辅助音标学习。

References 参考文献

[1] 鲁煊，文歧福，沈培奇. 中式烹调工艺：烹调技法训练 [M]. 成都：西南交通大学出版社，2021.

[2] 鲁煊，文歧福. 现代厨房管理实务 [M]. 北京：机械工业出版社，2021.

[3] 王艳玲. 烹饪英语 [M]. 北京：中国旅游出版社，2018.

[4] 北京市人民外事办公室，北京市民讲外语活动组委会办公室. 美食译苑：中文菜单英文译法 [M]. 北京：世界知识出版社，2011.

[5] 冯源. 简明中餐餐饮汉英双解辞典 [M]. 北京：北京大学出版社，2009.

[6] 何宏. 中外饮食文化 [M]. 2 版. 北京：北京大学出版社，2016.

[7] 吕尔欣. 中西方饮食文化差异及翻译研究 [M]. 杭州：浙江大学出版社，2013.

[8] 陈丕琮. 英汉餐饮词典 [M]. 上海：上海译文出版社，1995.

[9] 罗伯特·马杰尔. 厨房英语 [M]. 4 版. 陈亚丽，傅文漪，鄂丽娟，改编. 北京：旅游教育出版社，2017.

[10] Child J, Bertholle L, Beck S. Mastering the Art of French Cooking[M]. New York: Knopf Publishing Group, 2001.